Table of Contents

W0013354

Introduction

In the intricate tapestry of human health and longevity, the gut emerges as a pivotal player, orchestrating a symphony of biochemical interactions that extend far beyond digestion. The link between gut health and longevity has become a subject of profound scientific inquiry, captivating researchers and health enthusiasts alike. This connection is not a mere coincidence but a testament to the intricate web of relationships governing the human body.

The human gut, often called the gastrointestinal tract, is an ecosystem teeming with trillions of microorganisms, including bacteria, viruses, fungi, and archaea. This diverse community, collectively known as the gut microbiota, coexists harmoniously with the human host, influencing various aspects of physiology, metabolism, and immune function. Recent scientific revelations have brought the gut microbiota into the spotlight, showcasing its role in digestive processes and shaping broader aspects of health and, importantly, lifespan.

The gut microbiota is a dynamic and diverse community that varies from person to person and is influenced by genetics, diet, lifestyle, and environmental exposures. Comprising trillions of microorganisms, the gut microbiota plays a crucial role in breaking down complex carbohydrates, synthesizing vitamins, and modulating the immune system. The composition and diversity of the gut microbiota are increasingly recognized as key determinants of overall health and, by extension, longevity.

Research has revealed a delicate balance within the gut microbial community, where the abundance of specific bacterial species can have profound health implications. The presence of beneficial bacteria, such as Bifidobacteria and Lactobacilli, is associated with improved nutrient absorption, enhanced immune function, and a reduced risk of inflammatory conditions. Conversely, an imbalance, known as dysbiosis, characterized by an overgrowth of harmful bacteria, is linked to various health issues, including gastrointestinal disorders, metabolic dysfunction, and chronic inflammation.

The communication network between the gut and the brain, known as the gut-brain axis, serves as a bidirectional conduit through which signals are exchanged. Emerging research suggests that this axis not only influences mood and mental well-being but also plays a role in cognitive function and neurodegenerative diseases. The gut microbiota, acting as a mediator in this intricate dialogue, produces metabolites and neurotransmitters that can impact the central nervous system.

The concept of the gut-brain axis challenges traditional views that the brain operates independently of the rest of the body. Instead, it highlights the profound interconnectivity between the gut and the brain, with implications for mental health and longevity. Studies have demonstrated associations between gut microbiota composition and neurological conditions such as Alzheimer's disease and Parkinson's disease, suggesting that maintaining a healthy gut may contribute to cognitive resilience in ageing.

Chronic inflammation is a hallmark of ageing and is implicated in the pathogenesis of numerous age-related diseases. With its vast immune system and intimate connection to the external environment, the gut emerges as a critical player in modulating inflammatory processes throughout the body. Dysbiosis in the gut microbiota has been linked to increased systemic inflammation, providing a potential mechanistic link between gut health and longevity.

Understanding the role of the gut in inflammation requires delving into the intricate interactions between the gut microbiota, the intestinal barrier, and the immune system. A healthy gut microbiota maintains a robust intestinal barrier, preventing the leakage of harmful substances into the bloodstream and mitigating inflammatory responses. Conversely, disruptions in this delicate balance can lead to chronic inflammation, accelerating the ageing process and predisposing individuals to age-related diseases.

The gut is not merely a passive absorber of nutrients but an active participant in the metabolism of dietary components. Recent studies have illuminated the role of the gut microbiota in nutrient metabolism, demonstrating its impact on energy extraction, nutrient absorption, and the bioavailability of essential compounds. This intricate interplay extends beyond immediate physiological effects, influencing long-term health outcomes and potentially contributing to lifespan modulation.

One key aspect of this gut-centric perspective on nutrient metabolism is the role of short-chain fatty acids (SCFAs), microbial metabolites produced during the fermentation of dietary fibers. SCFAs, such as butyrate, acetate, and propionate, have been shown to exert anti-inflammatory effects, regulate energy metabolism, and maintain intestinal barrier integrity. The production of SCFAs by gut bacteria influences local gut health and has systemic effects, impacting various organs and tissues throughout the body.

Dietary choices profoundly influence the composition and function of the gut microbiota, serving as a modifiable factor that individuals can leverage to promote longevity. The Western diet, characterized by high levels of processed foods, sugars, and saturated fats, has been associated with dysbiosis and increased inflammation, contributing to the development of chronic diseases that can shorten lifespan.

Conversely, diets rich in fiber, prebiotics, and fermented foods support a diverse and beneficial gut microbiota. The Mediterranean diet, for instance, has been lauded for its potential to promote gut health through consuming fruits, vegetables, whole grains, and probiotic-rich foods. Understanding the intricate relationship between diet, the gut microbiota, and longevity provides a foundation for personalized dietary interventions to optimize healthspan.

The burgeoning field of microbiome research has paved the way for innovative interventions designed to modulate the gut microbiota and, by extension, promote longevity. Probiotics, live microorganisms with purported health benefits, have gained popularity as supplements and functional foods. When introduced into the gut, these beneficial bacteria can augment the existing microbiota and contribute to a more balanced microbial community.

Prebiotics, on the other hand, are non-digestible dietary fibers that selectively nourish beneficial bacteria in the gut. By providing a substrate for the growth of specific microbial species, prebiotics can help restore and maintain a healthy gut microbiota. The synergy between probiotics and prebiotics, known as synbiotics, represents a promising avenue for interventions to optimize gut health and enhance longevity.

Beyond diet, various lifestyle factors influence the gut-longevity connection. Physical activity, for example, has been shown to modulate the composition of the gut microbiota, promoting diversity and the abundance of beneficial bacteria. Chronic stress, on the other hand, can negatively impact gut health, leading to dysbiosis and increased susceptibility to inflammatory conditions.

Understanding the role of lifestyle factors in shaping the gut-longevity connection underscores the importance of holistic approaches to health promotion. Integrating strategies encompassing dietary modifications, physical activity, stress management, and adequate sleep can collectively contribute to a resilient and balanced gut microbiota, fostering longevity and overall well-being.

In conclusion, the gut-longevity connection represents a captivating frontier in biomedical research, illuminating the profound impact of the gut microbiota on overall health and lifespan. As we delve deeper into the intricacies of this relationship, the potential for targeted interventions to enhance gut health and promote longevity becomes increasingly evident. From the microbial symphony within our intestines to the far-reaching implications on neurological health, inflammation, and nutrient metabolism, the gut emerges as a nexus of influence on aging.

As we embark on this journey of exploration, it is essential to recognize the dynamic nature of the gut-longevity connection and the multifaceted factors that contribute to its complexity. The quest to unveil the secrets held within the gut microbiota promises a deeper understanding of human biology and the potential to redefine the healthcare landscape, ushering in an era where personalized strategies for gut health optimization contribute to a longer and healthier life.

Chapter One

The Evolution of Gut Health

The human gut, a complex ecosystem of trillions of microorganisms, plays a pivotal role in our overall health and well-being. Understanding the evolution of gut health is essential for grasping the foundation of current practices and gaining insights into the intricate relationship between the gut and human health. In this chapter, we will journey through time, exploring the historical context of gut health and how our understanding has evolved over the years.

Ancient Perspectives on Gut Health

The roots of gut health awareness trace back to ancient civilizations, where various cultures recognized the significance of digestive well-being. Ancient Greek physicians, such as Hippocrates, emphasized the importance of a balanced diet and its impact on overall health. They observed the connection between digestive disorders and various illnesses, laying the groundwork for understanding gut health.

In traditional Chinese medicine and Ayurveda, the ancient medical systems of China and India, respectively, the focus on maintaining digestive balance has been a cornerstone for thousands of years. These early perspectives recognized the gut as a central hub for physical and mental health, underscoring the holistic nature of well-being.

The Middle Ages: Limited Understanding and Herbal Remedies

The Middle Ages marked a period of limited scientific progress, and the understanding of gut health was often intertwined with superstitions and folklore. However, herbal remedies and dietary practices were prominent during this time, with various concoctions and diets believed to promote digestive health. While only sometimes scientifically sound, these practices intuitively acknowledge the gut's role in maintaining health.

Renaissance and the Emergence of Anatomical Knowledge

The Renaissance era witnessed a revival of scientific inquiry, bringing a more systematic understanding of anatomy and physiology. Pioneering anatomists like

Andreas Vesalius dissected human cadavers, providing detailed insights into the structure of the digestive system. This period laid the groundwork for a more empirical approach to medicine as the link between gut anatomy and health became increasingly apparent.

The Germ Theory Revolutionizes Gut Health

The 19th century saw a paradigm shift in medicine with the acceptance of the germ theory of disease. Scientists like Louis Pasteur and Robert Koch elucidated the role of microorganisms in causing illnesses. This revelation profoundly impacted our understanding of gut health, highlighting the delicate balance between beneficial and harmful microbes in the digestive system.

The recognition of probiotics, beneficial bacteria that promote gut health, gained traction during this period. Fermented foods and yogurts became associated with positive digestive outcomes, laying the foundation for the modern emphasis on probiotic-rich diets.

20th Century: Antibiotics, Processed Foods, and Changing Diets

The 20th century brought unprecedented advancements in medicine and technology, influencing human health and dietary practices. The discovery of antibiotics revolutionized medicine and led to unintended consequences for gut health. Widespread use of antibiotics, while essential for treating infections, disrupted the delicate microbial balance in the gut, giving rise to new challenges in maintaining digestive well-being.

Simultaneously, the rise of processed foods and changes in dietary habits profoundly impacted gut health. High-sugar, low-fiber diets became more prevalent, contributing to a host of digestive issues. This era witnessed a growing disconnect between traditional dietary practices and the evolving landscape of gut health.

Contemporary Understanding: Microbiome and Holistic Health

The late 20th and early 21st centuries marked a resurgence of interest in gut health, fueled by technological advancements that allowed deeper microbiome exploration. The human microbiome, a diverse community of microorganisms residing in the gut, became a key player in maintaining overall health.

Advancements in DNA sequencing technology facilitated a comprehensive gut microbiome analysis, revealing its immense diversity and intricate relationship with various aspects of human health. This newfound understanding prompted a shift toward holistic health practices, prioritizing the symbiotic relationship between the host and its microbial inhabitants.

Present-Day Practices: Probiotics, Prebiotics, and Personalized Nutrition

In the contemporary era, integrating scientific knowledge has given rise to practical strategies for enhancing gut health. Probiotics, once associated with ancient fermentation processes, are now available in various forms, from supplements to specialized foods. These beneficial bacteria aim to restore and maintain a healthy microbial balance in the gut.

The importance of prebiotics, which nourishes beneficial microbes, has also gained recognition. A shift towards dietary fiber-rich foods and a diverse range of nutrients supports the flourishing of a balanced microbiome.

Personalized nutrition has emerged as a key trend, recognizing that individuals may respond differently to dietary interventions based on their unique microbiome composition. This approach tailors dietary recommendations to an individual's specific microbial profile, maximizing the potential for gut health optimization.

While significant strides have been made in understanding gut health, challenges persist. The impact of modern lifestyles, including stress, sedentary habits, and environmental factors, continues to affect the delicate balance of the gut microbiome. The overuse of antibiotics and the prevalence of processed foods pose ongoing threats to gut health.

Future directions in gut health research involve delving deeper into the intricacies of the microbiome and developing innovative interventions. The exploration of fecal microbiota transplantation (FMT), the transfer of healthy microbes from one individual to another, holds promise for addressing certain gastrointestinal conditions and further highlights the dynamic nature of gut health practices.

The evolution of gut health reflects the dynamic interplay between scientific advancements, cultural practices, and societal changes. From ancient wisdom to contemporary scientific insights, our understanding of gut health has come a long way. Recognizing the gut as a central player in overall health emphasizes the need for holistic approaches considering both the host and its microbial inhabitants.

As we navigate the complexities of modern life, integrating the lessons from the past with cutting-edge research can guide us toward practices that promote a balanced and resilient gut. This journey through the history of gut health underscores the ongoing quest to unravel the mysteries of the microbiome. It highlights the potential for transformative approaches to well-being in the years to come.

Chapter Two

Gut Microbiota Diversity - The Key to Resilience

The human gut, home to trillions of microorganisms collectively known as the gut microbiota, forms a complex ecosystem that profoundly influences our health. This chapter delves into the intricate world of gut microbiota diversity, recognizing it as a key determinant of resilience and well-being. Understanding the critical role of diverse gut microbiota is essential for shaping contemporary perspectives on health and disease.

The gut microbiota comprises many bacteria, viruses, fungi, and other microorganisms that coexist in a dynamic equilibrium. This microbial community, often called the "forgotten organ," contributes significantly to various physiological processes, including digestion, metabolism, and immune function. The diversity within this microbial tapestry is crucial in maintaining a balanced and resilient gut environment.

The Guardians of Gut Health: Bacterial Diversity

Bacterial diversity is a cornerstone of a healthy gut ecosystem. Different species of bacteria perform distinct functions, contributing to the breakdown of complex carbohydrates, synthesizing essential vitamins, and preventing harmful pathogens from establishing residence. The intricate balance between beneficial, neutral, and potentially harmful bacteria is a testament to the evolutionary coexistence that has shaped human-microbe relationships over millennia.

Microbiota and the Immune System

The gut microbiota plays a pivotal role in shaping the development and function of the immune system. Diverse microbial communities stimulate the immune system, training it to distinguish between friend and foe. This immunomodulatory role of the gut microbiota is essential for mounting effective responses against pathogens while maintaining tolerance to the body's tissues. A lack of microbial diversity may compromise the immune system's ability to navigate this delicate balance, potentially leading to autoimmune disorders and increased susceptibility to infections.

The Gut-Brain Axis: Microbial Influence on Mental Health

Beyond its impact on physical health, the gut microbiota profoundly influences mental well-being through the gut-brain axis. Communication between the gut and the central nervous system involves a complex interplay of biochemical signals, with the microbiota acting as a key mediator. Diverse microbial communities produce neurotransmitters and other signaling molecules influencing mood, cognition, and stress responses. Imbalances in gut microbiota diversity have been linked to mental health disorders, emphasizing the need to preserve this diversity for optimal brain-gut communication.

Disruptions to Microbiota Diversity: Antibiotics and Modern Lifestyles

While the human-microbe relationship has evolved over millennia, modern lifestyles and medical interventions have introduced novel challenges to gut microbiota diversity. Antibiotics, while life-saving in many instances, can indiscriminately deplete harmful and beneficial bacteria, disrupting the delicate balance within the gut.

This disturbance may have far-reaching consequences for health, including an increased risk of antibiotic-resistant infections and compromised gut function.

Additionally, factors such as a high-sugar, low-fiber diet, sedentary habits, and chronic stress decrease microbial diversity. These elements of contemporary living have created an environment that is less conducive to nurturing robust and varied gut microbiota. Recognizing these challenges is paramount for developing strategies that promote diversity in the face of modern pressures.

The Significance of Preserving and Enhancing Microbiota Diversity

1. **Resilience to Pathogens:** Diverse gut microbiota acts as a natural barrier against pathogenic invaders. A rich microbial community occupies ecological niches, leaving little room for harmful bacteria to establish themselves. This microbial competition and the production of antimicrobial substances contribute to a resilient defense against infections.

2. **Metabolic Health and Weight Regulation:** Microbiota diversity has been linked to metabolic health and weight regulation. Certain bacterial species are associated with efficient energy

extraction from food, and an imbalance in microbial diversity may contribute to obesity and metabolic disorders. Preserving a diverse gut microbiota may play a role in maintaining a healthy weight and metabolic profile.

3. **Digestive Harmony:** Diverse gut microbiota is essential for maintaining digestive harmony. Different bacterial species contribute to the breakdown of various components of our diet, aiding in the absorption of nutrients and preventing digestive disorders. Imbalances in microbiota diversity may lead to conditions such as irritable bowel syndrome (IBS) and inflammatory bowel diseases (IBD).

4. **Immune System Fitness:** A diverse gut microbiota stimulates the immune system, helping it distinguish between harmful pathogens and harmless substances. This immunomodulatory role is crucial for preventing autoimmune disorders and enhancing the body's ability to mount effective immune responses when needed.

5. **Mental Well-Being:** The gut-brain axis, influenced by diverse gut microbiota, is vital to mental well-being. Maintaining microbial diversity is associated

with a lower risk of mood disorders, anxiety, and depression. Strategies that preserve and enhance gut microbiota diversity may improve mental health outcomes.

Strategies for Preserving and Enhancing Microbiota Diversity

1. **Dietary Interventions:** A diet rich in diverse fibers, polyphenols, and fermented foods supports microbial diversity. These components serve as prebiotics, nourishing beneficial bacteria and promoting a balanced microbial ecosystem. Emphasizing whole foods and reducing the intake of processed and sugary foods is integral to preserving gut microbiota diversity.

2. **Probiotics and Fermented Foods**: Supplementing with probiotics and incorporating fermented foods into the diet introduces beneficial bacteria, contributing to microbial diversity. Yogurt, kefir, sauerkraut, and kimchi are examples of fermented foods that have been associated with positive effects on gut health.

3. **Avoiding Unnecessary Antibiotic Use:** Prudent use of antibiotics is essential to prevent unnecessary disruptions to gut microbiota diversity. Healthcare providers and individuals alike play a role in ensuring that antibiotics are prescribed judiciously, considering the potential impact on microbial balance.

4. **Lifestyle Modifications:** Regular physical activity, managing stress through meditation and mindfulness, and getting adequate sleep contribute to a healthy gut environment. These lifestyle modifications support overall well-being and help maintain microbial diversity.

5. **Personalized Approaches:** Recognizing individual variations in gut microbiota composition, personalized approaches to nutrition and health are gaining prominence. Advances in microbiome research enable the tailoring of interventions to an individual's unique microbial profile, optimizing the preservation and enhancement of gut microbiota diversity.

Despite the growing awareness of the importance of gut microbiota diversity, challenges persist in implementing strategies for its preservation. Societal factors, such as the prevalence of processed foods and the overuse of antibiotics, pose ongoing threats. Additionally, the complexity of the gut microbiome and its interactions with various aspects of health necessitate continued research to uncover nuanced insights.

Future directions in gut microbiota research involve unraveling the specific mechanisms underlying the link between microbial diversity and health outcomes. This understanding may pave the way for targeted interventions that address individual needs, optimizing gut health in a personalized manner.

The previous chapter has illuminated the critical role of diverse gut microbiota in maintaining overall health and resilience. The microbial communities within our gut are integral to various physiological processes, influencing everything from immune function to mental well-being. Recognizing, preserving, and enhancing gut microbiota diversity is a cornerstone of contemporary health practices.

As we navigate the challenges of modern living, strategies that promote microbial diversity, such as dietary interventions, probiotics, and lifestyle modifications, become paramount. The ongoing quest to understand the intricacies of the gut microbiome and its dynamic relationship with human health underscores the evolving nature of our approach to well-being. By embracing the lessons from the last chapter, we move one step closer to unlocking the full potential of our symbiotic relationship with the microbial world within.

Chapter Two

Gut Microbiota Diversity - A Vital Key to Resilience

In the journey of ageing, maintaining good health becomes a paramount focus. This chapter delves into the intricate world of gut microbiota diversity, recognizing it as a key determinant of resilience and well-being tailored to the unique needs of individuals. Understanding the critical role of diverse gut microbiota is essential for shaping contemporary perspectives on health and addressing the particular challenges faced during this stage of life.

The Microbial Tapestry Within: Tailored for Age and Experience

As individuals progress into their senior years, the intricate microbial tapestry within the gut undergoes a nuanced transformation, reflective of the age and experience amassed over a lifetime. This biological masterpiece, comprising trillions of microorganisms, assumes an even more pivotal role in shaping health outcomes during this distinct phase of life.

1. **The Symphony of Experience:** Like an intricately composed symphony, the gut microbiota of individuals over 60 reflects the diverse experiences, environments, and dietary choices encountered throughout a lifetime. This rich history imprints itself on the microbial communities, creating a unique and personalized composition that distinguishes the gut of a senior individual from that of their younger counterparts.

2. **Evolutionary Harmony:** The aging gut microbiota represents the culmination of an evolutionary dance between the host and its microbial inhabitants. This symbiotic relationship, forged over decades, adapts to the changing physiological landscape of the aging body. The microbial players, finely tuned to the specific needs and challenges of aging, work in harmony to support various bodily functions.

3. **Adaptation to Nutritional Shifts:** Years of dietary habits and nutritional choices leave an indelible mark on the gut microbiota. As individuals age, the microbial communities within the gut adapt to shifts in nutritional requirements. The tailored response of the microbiota becomes especially crucial, ensuring efficient nutrient extraction and absorption to support the changing needs of the aging body.

4. **Resilience Forged Through Challenges**: The microbial tapestry within the gut of individuals over 60 reflects a remarkable resilience forged through a lifetime of challenges. From encounters with various pathogens to exposures to diverse environments, the gut microbiota adapts and fortifies its defenses, contributing to the overall resilience of the individual.

5. **Influence of Lifestyle and Stress:** The impact of lifestyle and stress on the gut microbiota becomes increasingly pronounced in the senior years. Decades of exposure to stressors and lifestyle choices shape the microbial communities. Strategies to preserve and enhance gut microbiota diversity during this phase must consider these cumulative influences, acknowledging the unique imprint of a lifetime's experiences.

6. **A Unique Microbial Identity:** The gut microbiota of individuals over 60 develops a unique microbial identity akin to a fingerprint of health. This personalized composition influences how the body responds to challenges, maintains homeostasis, and supports overall well-being. Understanding and respecting this individualized microbial identity is fundamental to crafting effective strategies for promoting gut health in the senior years.

Implications for Health and Well-being

Understanding the microbial tapestry within the aging gut carries profound implications for health and well-being. The personalized nature of the gut microbiota in individuals underscores the importance of tailored interventions that acknowledge and leverage the unique characteristics of their microbial communities.

1. **Targeted Nutrition:** Tailored nutrition becomes a key consideration when recognizing the adaptability of the gut microbiota to nutritional shifts. Senior individuals can benefit from dietary approaches that align with their unique microbial composition, ensuring optimal nutrient absorption and supporting overall health.

2. **Stress Management and Lifestyle Interventions:** Acknowledging the cumulative impact of stress and lifestyle on the gut microbiota calls for targeted interventions in stress management and lifestyle modifications. Practices that promote relaxation, regular physical activity, and healthy sleep contribute to overall well-being and the preservation of microbial diversity.

3. **Personalized Health Plans:** The individualized microbial identity of the aging gut warrants personalized health plans. Healthcare providers can leverage advancements in microbiome research to tailor interventions, medications, and health recommendations to each individual's specific needs and characteristics, optimizing health outcomes.

4. **Holistic Approaches:** Embracing a holistic approach to health that integrates physical, mental, and microbial well-being becomes paramount. Viewing health through the lens of the unique microbial tapestry within the gut encourages comprehensive strategies that address the multifaceted nature of health during the senior years.

Exploring the microbial tapestry within the aging gut extends beyond mere observation; it opens avenues for practical applications and future research. As science advances, the following considerations emerge:

1. **Precision Medicine in Geriatrics:** Advancements in microbiome research pave the way for precision medicine tailored specifically for geriatric populations. Tailoring medical interventions, medications, and health plans based on an individual's microbial profile

holds immense potential for optimizing health outcomes in the senior years.

2. **Innovative Dietary Approaches:** Future research may uncover innovative dietary approaches that leverage the unique microbial composition of individuals over 60. Personalized dietary plans, rich in prebiotics and tailored to the specific needs of the aging gut, could become integral components of health strategies for the senior population.

3. **Microbial-Based Therapies:** Exploration into microbial-based therapies, including precision probiotics and microbial transplants, holds promise for addressing age-related health challenges. These therapies, informed by a deep understanding of the aging gut microbiota, may offer targeted solutions for promoting resilience and well-being.

4. **Longitudinal Studies on Aging Microbiota:** Longitudinal studies tracking the changes in the gut microbiota over the entire lifespan contribute valuable insights. Understanding how the microbial tapestry evolves from youth to old age enables the development of interventions that support healthy aging and prevent age-related diseases.

The microbial tapestry within the gut, tailored for age and experience, paints a vivid portrait of health in individuals over 60. This unique microbial journey, shaped by a lifetime of experiences, adaptations, and challenges, underscores the importance of personalized approaches to health. By embracing the distinctive characteristics of the aging gut microbiota, individuals can embark on a journey of well-being that respects the intricacies of their microbial tapestry and promotes health in the context of a life well-lived.

The Guardians of Gut Health for Seasoned Individuals: Bacterial Diversity

As individuals gracefully age into their seasoned years, the guardianship of gut health becomes increasingly entrusted to the intricate dance of bacterial diversity within. These bacterial custodians play a pivotal role in shaping the landscape of the gut microbiota, orchestrating a symphony of interactions that influence digestive processes and the broader tapestry of health and resilience.

1. **The Diverse Cast of Bacterial Characters:** Bacterial diversity within the gut of seasoned individuals resembles a diverse cast of characters, each with unique strengths

and contributions. These bacteria, ranging from Firmicutes to Bacteroidetes and beyond, create a harmonious ecosystem that fosters a delicate balance between stability and adaptability. This intricate balance becomes a cornerstone for maintaining gut health and overall well-being.

2. **Digestive Maestros:** The bacterial diversity in the gut serves as a legion of digestive maestros, specializing in the breakdown of complex compounds into nutrients that the body can absorb and utilize. This becomes particularly crucial in the seasoned years when the efficiency of nutrient absorption may face challenges. The diverse bacterial ensemble ensures that the digestive symphony continues to play with finesse, contributing to optimal nutritional status for seniors.

3. **Synthesizers of Essential Harmony:** Certain bacterial cohorts within the gut microbiota act as synthesizers of essential harmony. They contribute to producing vital substances such as vitamins, short-chain fatty acids, and other bioactive compounds. This microbial synthesis becomes instrumental in supporting metabolic processes, immune function, and overall health, offering a symphony of support for seasoned individuals.

4. **Guardians Against Invaders:** Bacterial diversity operates as a formidable force, acting as a guardians against potential invaders that could disrupt the delicate balance of the gut ecosystem. A diverse bacterial community creates a crowded environment, leaving limited room for harmful pathogens to establish themselves. This natural defense mechanism becomes increasingly valuable as the immune system may experience shifts with age.

5. **Balancers of Inflammation:** Certain bacterial species within the diverse gut microbiota act as balancers of inflammation. Inflammation, a natural response to challenges, can become dysregulated with age. The balanced presence of anti-inflammatory and pro-inflammatory bacteria within the gut contributes to the nuanced regulation of inflammation, fostering an environment that supports immune function without tipping the scales into chronic inflammation.

6. **Symbiotic Relationships with the Host:** The bacterial diversity within the seasoned gut engages in symbiotic relationships with the host. This intricate dance involves the exchange of signals, nutrients, and support. The bacteria receive a resource-rich habitat, while the host benefits from the microbial contributions to digestion,

metabolism, and immune function. This symbiotic relationship is a testament to the coevolutionary journey shared by humans and their gut bacteria.

7. **Adaptability to Dietary Shifts:** Seasoned individuals often experience shifts in dietary preferences and nutritional needs. The bacterial diversity within the gut showcases remarkable adaptability to these changes. This adaptability ensures that the microbial community can efficiently process and derive benefits from a diverse range of foods, contributing to the resilience of the gut ecosystem.

8. **Guardians of Mental Well-Being:** Beyond physical health, the diverse gut microbiota of seasoned individuals acts as guardians of mental well-being. The microbial inhabitants influence the gut-brain axis, a bidirectional communication system between the gut and the central nervous system. Bacterial diversity contributes to the production of neurotransmitters and other signaling molecules, influencing mood, cognition, and stress responses.

Strategies to Support Bacterial Diversity in Seasoned Years

Acknowledging the vital role of bacterial diversity in maintaining gut health for seasoned individuals prompts exploring strategies to support and enhance this diversity.

1. **Dietary Richness:** A diet rich in diverse fibers, prebiotics, and a spectrum of nutrients nourishes a varied microbial community. Embracing a colorful array of fruits, vegetables, whole grains, and legumes becomes a cornerstone for fostering bacterial diversity.

2. **Probiotics and Fermented Foods:** Incorporating probiotics and fermented foods introduces beneficial bacteria, contributing to microbial balance. Yogurt, kefir, sauerkraut, and kimchi are examples of fermented foods that can enhance bacterial diversity and support gut health.

3. **Mindful Antibiotic Use:** Prudent and mindful use of antibiotics becomes crucial for preserving bacterial diversity. Seasoned individuals, in consultation with healthcare providers, should consider alternatives and only resort to antibiotics when necessary, minimizing potential disruptions to the gut microbiota.

4. **Active Lifestyle:** Regular physical activity is not only beneficial for overall health but also supports a healthy gut environment. Exercise has been associated with positive changes in the gut microbiota, contributing to diversity and resilience.

5. **Stress Management:** Given the impact of stress on the gut-brain axis and microbial balance, stress management practices such as meditation, mindfulness, and relaxation techniques become integral components of a holistic approach to gut health for seasoned individuals.

Looking Ahead: Unveiling the Mysteries of Bacterial Diversity

The guardianship of gut health for seasoned individuals by bacterial diversity unveils a realm of mysteries that continues to captivate researchers and health enthusiasts alike. As science advances, understanding specific bacterial strains, their roles, and their interplay within the gut ecosystem will likely deepen, opening avenues for even more targeted and personalized strategies to support gut health in the seasoned years.

The guardianship of gut health for seasoned individuals orchestrated by bacterial diversity is a testament to the intricate dance between the microbial world and the human host. Nurturing these guardians within involves embracing dietary, lifestyle, and healthcare practices that respect the delicate balance of the gut ecosystem. As we navigate the seasoned years, the wisdom of fostering bacterial diversity becomes a key ally in the journey toward continued health, vitality, and well-being.

Microbiota and the Aging Immune System: A Symphony of Influence

As individuals gracefully traverse the aging journey, the interplay between the microbiota and the immune system takes center stage, influencing overall health harmony.

This intricate relationship becomes increasingly crucial as the aging immune system encounters unique challenges and opportunities.

Understanding the nuanced dynamics between microbiota and immune function sheds light on strategies to promote resilient health in the golden years.

The Dynamic Landscape of the Aging Immune System

The ageing process orchestrates a symphony of changes within the immune system. Immunosenescence, the gradual decline in immune function, is accompanied by alterations in both the innate and adaptive arms of immunity. Key players, such as T cells and B cells, exhibit changes in functionality and responsiveness, impacting the body's ability to mount effective immune responses.

Gut Microbiota as an Architect of Immune Balance

The gut microbiota emerges as a chief architect in maintaining immune balance during aging. This microbial community, residing within the gastrointestinal tract, interacts intimately with the immune cells, shaping their development, function, and responsiveness. The symbiotic relationship between the microbiota and the immune system is a dynamic exchange that has far-reaching implications for health in the senior years.

Immunomodulation by Gut Microbes

Gut microbes actively engage in immunomodulation, influencing the delicate balance between inflammatory and anti-inflammatory responses. In the context of ageing, where chronic low-grade inflammation, termed "inflammaging," is a hallmark, the role of the microbiota becomes pivotal. Specific microbial species regulate inflammation, aiming for controlled immune activation without excessive, detrimental inflammation.

The Thymus and T Cell Dynamics

The thymus, a critical organ for T cell development, undergoes involution with age, impacting the diversity and functionality of T cells. The microbiota plays a nuanced role in shaping T-cell dynamics. Microbial signals influence the maturation of T cells in the thymus, contributing to the maintenance of a diverse and functional T cell repertoire.

Gut Microbiota and Immune Memory

The concept of immune memory, essential for mounting rapid and effective responses to previously encountered pathogens, extends beyond adaptive immunity. The gut microbiota contributes to the training of innate immune cells, such as macrophages and dendritic cells, promoting heightened alertness and responsiveness. This training effect, known as trained immunity, has implications for the defense against infections and the overall robustness of the immune system.

Impact on Vaccination Responses

Vaccination, a cornerstone of preventive medicine, becomes a focal point in the context of the aging immune system. The gut microbiota influences the efficacy of vaccinations by shaping the body's response to immunization. Understanding how specific microbial communities enhance or dampen vaccine responses is crucial for optimizing vaccination strategies in older people.

Gut Dysbiosis and Immune Dysfunction

Dysbiosis, an imbalance in the composition and function of the gut microbiota, is common in the aging gut. This dysbiosis is associated with alterations in immune function, contributing to increased susceptibility to infections and a higher risk of inflammatory conditions. Restoring microbial balance becomes a potential avenue for mitigating immune dysfunction in older people.

Strategies to Support Microbiota-Immune Harmony in Aging

Recognizing the symbiotic relationship between the gut microbiota and the aging immune system prompts exploring strategies to foster harmony and support immune resilience.

1. **Prebiotics and Probiotics:** Including prebiotics, which nourishes beneficial microbes, and probiotics, which introduce live beneficial bacteria, in the diet supports the maintenance of healthy and diverse gut microbiota. These interventions have been linked to improvements in immune function in older people.

2. **Dietary Diversity:** A diet rich in diverse nutrients, including antioxidants and anti-inflammatory compounds, supports the gut microbiota and immune system. Emphasizing a colorful array of fruits, vegetables, whole grains, and lean proteins contributes to overall health and resilience.

3. **Physical Activity:** Regular physical activity has been associated with positive effects on the gut microbiota and the immune system. Exercise promotes microbial

diversity and enhances immune function, providing a holistic approach to supporting health in the aging population.

4. **Adequate Sleep:** Quality sleep is integral to immune function and microbial balance. Prioritizing adequate sleep supports the body's ability to regulate immune responses and maintain a healthy gut environment.

5. **Mindful Antibiotic Use:** Prudent antibiotic use, carefully considering potential impacts on the gut microbiota, helps prevent unnecessary disruptions. Mindful antibiotic management becomes particularly relevant for preserving immune-microbiota balance in older people.

Unlocking Potential Therapies

The evolving understanding of the interplay between the gut microbiota and the aging immune system opens doors to potential therapeutic interventions. Future research may unveil targeted approaches, including precision probiotics, microbial-based therapies, and personalized immunomodulatory strategies, designed to enhance immune resilience in older people.

The symbiotic dance between the gut microbiota and the aging immune system orchestrates a symphony of influence that shapes health outcomes in the senior years. Navigating this intricate symphony involves:

- Recognizing the nuanced connections.
- Embracing strategies that support microbial and immune harmony.
- Fostering an environment that promotes resilience.

As we unveil the mysteries of this symbiotic relationship, we pave the way for a healthier and more vibrant journey through the symphony of aging immunity.

The Gut-Brain Axis: Nurturing Mental Well-Being

As individuals enter the golden years, the intricate communication highway known as the gut-brain axis assumes even greater significance in shaping mental well-being. This bidirectional relationship between the gut and the brain orchestrates a complex interplay of signals, neurotransmitters, and microbial influences that extend far beyond the digestive realm. Understanding and nurturing the gut-brain axis is key to promoting cognitive vitality and emotional resilience in the senior years.

1. The Neurotransmitter Orchestra: The gut, often called the "second brain," houses a vast network of neurons called the enteric nervous system. This network communicates with the central nervous system, creating a symphony of neurotransmitters. Serotonin, dopamine, and gamma-aminobutyric acid (GABA) are key players in this neurotransmitter orchestra. These molecules influence mood and emotions and contribute to cognitive function, memory, and overall mental well-being.

2. Microbial Composers: Microbes residing in the gut actively participate in composing the gut-brain melody. The gut microbiota produces neurotransmitters and neuroactive compounds that can directly influence the enteric nervous system. For example, certain bacteria generate short-chain fatty acids (SCFAs), which have been linked to neuroprotective effects and mood regulation.

3. Impact on Cognitive Function: The gut-brain axis profoundly impacts cognitive function, a consideration of paramount importance in the golden years. Research suggests that disruptions in the gut microbiota composition can be associated with cognitive decline and neurodegenerative diseases. Conversely, a balanced and diverse gut microbiota contributes to cognitive resilience, influencing learning, memory, and information processing.

4. Emotional Resilience and Mood Regulation: The gut-brain axis is a key regulator of emotional resilience and mood. The communication between the gut and the brain influences responses to stress, anxiety, and depression. Alterations in the gut microbiota composition have been linked to mood disorders, emphasizing the need to prioritize gut health for mental well-being in the senior years.

5. Inflammation and Neurological Health: Chronic inflammation, often associated with age-related conditions, can affect the gut and the brain. The gut-brain axis plays a crucial role in regulating inflammation, and disruptions in this axis may contribute to neuroinflammation. Managing inflammation through a balanced gut environment becomes crucial for preserving neurological health in the golden years.

6. Nutrient Production and Brain Health: The gut microbiota actively produces essential nutrients, some of which play a vital role in brain health. For instance, certain bacteria contribute to the synthesis of B vitamins, which are integral for cognitive function. Maintaining a healthy gut environment ensures the availability of these essential nutrients to support brain health.

Strategies for Nurturing the Gut-Brain Axis

Understanding the intricate connection between the gut and the brain prompts the exploration of strategies to nurture the gut-brain axis for mental well-being:

1. **Probiotics and Fermented Foods:** Incorporating probiotics and fermented foods into the diet introduces beneficial bacteria that can positively influence the gut-brain axis. Yogurt, kefir, kimchi, and sauerkraut are examples of foods that support a healthy microbial balance.

2. **Prebiotics for Microbial Nourishment:** Prebiotic-rich foods, such as garlic, onions, leeks, and asparagus, nourish beneficial microbes. These foods promote the growth of microbes that contribute to neurotransmitter production and overall gut health.

3. **Dietary Omega-3 Fatty Acids:** Omega-3 fatty acids in fatty fish like salmon and flaxseeds have anti-inflammatory properties and support cognitive function.

Including these foods in the diet contributes to both gut and brain health.

4. **Mindful Eating Practices:** Adopting mindful eating practices, such as chewing food thoroughly and savoring each bite, supports optimal digestion. The mind-body connection during meals contributes to a positive gut-brain axis environment.

5. **Regular Physical Activity:** Physical activity has been linked to a healthy gut microbiota and improved cognitive function. Regular exercise promotes a favorable gut-brain axis environment, benefiting mental and physical well-being.

6. **Adequate Sleep:** Quality sleep is essential for cognitive health and emotional well-being. Prioritizing adequate sleep supports the regulation of neurotransmitters and contributes to a balanced gut-brain axis.

7. **Stress Management Techniques:** Practicing stress management techniques, such as meditation, deep breathing, and yoga, supports a balanced gut-brain axis. Chronic stress can negatively impact the gut microbiota and contribute to mental health challenges.

As our understanding of the gut-brain axis advances, future frontiers in research may unveil precision strategies tailored to individual needs. Personalized approaches, including targeted dietary interventions, probiotic formulations, and lifestyle recommendations, promise to optimize cognitive health in the golden years.

Nurturing the gut-brain axis emerges as a harmonious endeavor for promoting mental well-being in the golden years. The symphony of neurotransmitters, microbial influences, and intricate communication between the gut and the brain underscores the interconnectedness of physical and mental health. By prioritizing strategies that support a balanced gut-brain axis, individuals can embark on a journey toward cognitive vitality, emotional resilience, and a fulfilling experience of their golden years.

Disruptions to Microbiota Diversity: Medications and Aging

As individuals age, the intricate balance of the gut microbiota, a thriving community of microorganisms in the digestive system, can face disruptions due to various factors.

Among these, medications play a significant role in altering the composition and diversity of the microbial landscape. Understanding the impact of medications on microbiota diversity becomes crucial for addressing the unique challenges posed to gut health during aging.

1. **Antibiotics; The Double-Edged Sword:** Antibiotics, while invaluable in treating bacterial infections, represent a double-edged sword for gut health. These medications, designed to target and eliminate harmful bacteria, can inadvertently affect beneficial microbes. The indiscriminate nature of antibiotics may lead to a temporary or even prolonged disruption in microbiota diversity, potentially compromising the overall balance of the gut ecosystem.

2. **Non-Steroidal Anti-Inflammatory Drugs (NSAIDs):** Balancing Act: Non-steroidal anti-inflammatory Drugs (NSAIDs), commonly used for pain management and inflammation, can exert effects on the gut mucosa and alter microbial populations. Chronic use of NSAIDs may contribute to changes in the gut environment, impacting microbial diversity. Achieving a delicate balance between managing pain and preserving gut health becomes a consideration for individuals relying on NSAIDs, particularly in older people.

3. **Proton Pump Inhibitors (PPIs); Altered Acidity, Altered Microbiota:** Proton Pump Inhibitors (PPIs) reduce stomach acidity and are frequently prescribed for managing acid reflux and peptic ulcers. This alteration in the gastric environment can influence the survival and

transit of microbes through the digestive tract. Prolonged use of PPIs may contribute to shifts in microbiota composition, highlighting the need for reasonable use and regular monitoring, especially in the aging population.

4. **Corticosteroids; Immune Modulation and Microbial Impact:** Corticosteroids, powerful anti-inflammatory agents, modulate the immune response but may also impact the gut microbiota. These medications can alter the balance between beneficial and harmful microbes, potentially leading to dysbiosis. The immunosuppressive effects of corticosteroids underscore the importance of cautious use and proactive measures to support gut health in aging individuals.

5. **Polypharmacy:** The phenomenon of polypharmacy, the concurrent use of multiple medications, is prevalent in the aging population. The cumulative impact of various medications on the gut microbiota becomes a complex interplay that can amplify disruptions. Managing polypharmacy and considering the collective effects of medications on gut health is essential for preserving microbiota diversity in the elderly.

6. **Aging Itself: A Mediator of Microbial Changes:** The ageing process itself introduces changes to the gut

microbiota. Factors such as alterations in immune function, changes in dietary habits, and a decline in physiological resilience can contribute to shifts in microbial diversity. Understanding the interplay between aging and medication-induced disruptions provides a holistic view of the challenges the gut microbiota faces in the senior years.

Strategies for Preserving Microbiota Diversity:

Recognizing the potential disruptions to microbiota diversity, especially in the context of medications and aging, prompts exploring strategies to mitigate these effects and promote a healthy gut environment.

1. Probiotics and Prebiotics: Integrating probiotics, which are beneficial live bacteria, and prebiotics, which nourish these bacteria, into the diet can support microbial balance. Probiotic-rich foods and prebiotic fibres help replenish and foster the growth of beneficial microbes, countering the impact of medications.

2. Dietary Diversity: Maintaining a diverse and nutrient-rich diet contributes to overall gut health. A wide range of fruits, vegetables, whole grains, and fibre sources provides essential nutrients and promotes microbial diversity.

3. Timed Medication Intake: When feasible, considering the timing of medication intake about meals and circadian rhythms may influence the impact on the gut microbiota. Coordinating medication schedules with natural digestive cycles can optimize the absorption and metabolism of medications.

4. Periodic Gut Health Assessments: Regular assessments of gut health, including microbial composition analysis, can provide insights into changes over time. Monitoring microbiota diversity allows for proactive interventions and adjustments in medications or lifestyle to preserve gut health.

5. Collaboration with Healthcare Providers: Open communication with healthcare providers is crucial, especially for seniors managing multiple medications. Discussing concerns about potential impacts on gut health and exploring alternative options or supportive measures ensures a balanced approach to overall well-being.

As the field of personalized medicine advances, future approaches to medication management may incorporate individualized considerations for gut health. Tailoring medications based on an individual's microbial profile and health status holds promise for minimizing disruptions and optimizing therapeutic outcomes.

The Significance of Preserving and Enhancing Microbiota Diversity During Aging

1. Resilience to Age-Related Conditions: Diverse gut microbiota acts as a natural defense against age-related conditions. A robust microbial community contributes to the body's resilience, helping fend off diseases and maintaining vitality in the golden years.

2. Supporting Nutrient Absorption: As the body ages, nutrient absorption can become less efficient. Diverse gut microbiota plays a role in breaking down and extracting essential nutrients from food, supporting overall nutritional well-being during the senior years.

3. Cognitive Health and Mental Resilience: Preserving diverse gut microbiota is key to maintaining cognitive health and mental resilience. Influenced by microbial diversity, the gut-brain axis is crucial in preventing age-related cognitive decline and supporting emotional well-being.

4. Immune Vigilance: For those in their senior years, supporting the immune system is paramount. Diverse gut microbiota contributes to immune vigilance, helping defend against infections and supporting the body's ability to respond to health challenges.

5. Digestive Harmony: Maintaining diverse gut microbiota is essential for digestive harmony. It helps prevent common gastrointestinal issues and contributes to overall digestive comfort, a significant consideration for those in their senior years.

Strategies Tailored for Aging Well

1. Nutrient-dense, Senior-Friendly Diets: Adopting a nutrient-dense diet that caters to specific nutritional needs during the senior years supports gut health. Emphasizing whole foods, fiber, and essential nutrients becomes a cornerstone for preserving microbial diversity.

2. Probiotics and Gut-Friendly Foods: Incorporating probiotics and gut-friendly foods into the diet becomes increasingly important. Yogurt, kefir, and fermented foods provide essential nutrients and introduce beneficial bacteria, supporting the microbial balance within the gut.

3. Mindful Medication Management: For those in their senior years, managing medications mindfully is crucial. Understanding the potential impacts of medications on gut health and working closely with healthcare providers to mitigate disruptions ensures a balanced approach to overall well-being.

4. Lifestyle Practices for Aging Well: Regular physical activity, managing stress through relaxation techniques, and prioritizing adequate sleep contribute to a healthy gut environment. These lifestyle practices become essential components of a holistic approach to health during the senior years.

5. Regular Gut Health Check-ins: Regular health check-ins that include assessments of gut health can provide valuable insights. Monitoring changes and addressing concerns promptly ensures proactive management of gut microbiota diversity during the senior years.

Chapter Three

Beyond Probiotics - Unlocking the Power of Prebiotics

The previous chapter has shed light on the critical role of diverse gut microbiota tailored for individuals in their senior years. The microbial communities within the gut, intricately linked to various aspects of health, become vital allies in navigating the challenges of ageing. By recognizing the significance of preserving and enhancing gut microbiota diversity, individuals in their golden years can embark on a journey toward sustained well-being, resilience, and an enhanced quality of life.

In gut health and microbiota management, the spotlight often shines on probiotics, those beneficial microorganisms that confer various health benefits. However, a holistic understanding of gut health extends beyond probiotics to embrace the crucial role of prebiotics. This chapter delves into the intricate world of prebiotics, exploring their unique contributions to fostering a resilient and diverse gut microbiota. As we unlock the power of prebiotics, we uncover a wealth of insights that pave the way for personalized strategies promoting optimal gut health.

The Essence of Prebiotics

The term "prebiotic" refers to substances that nourish beneficial microorganisms residing in the gut. Unlike probiotics, which are live microorganisms, prebiotics are non-digestible fibers and compounds that pass through the digestive tract intact, reaching the colon, where they become a source of sustenance for specific microbes.

The Fiber Connection

Prebiotics are often synonymous with dietary fibers, and many prebiotics fall within the fiber category. Dietary fibres, encompassing soluble and insoluble varieties, serve as a buffet for the gut microbiota. Soluble fibers in foods like oats, legumes, and fruits undergo fermentation in the colon, producing short-chain fatty acids (SCFAs) that nourish the gut lining and support microbial diversity.

Beyond Fibers

While fibres are prominent prebiotics, other compounds also qualify for this essential role. Inulin, a fructan found in foods like chicory root and garlic, and oligosaccharides in onions and leeks exemplify non-fiber prebiotics. These compounds resist digestion in the upper gastrointestinal tract, reaching the colon intact to fuel the growth of beneficial bacteria.

Prebiotics and the Gut Microbiota

The relationship between prebiotics and the gut microbiota is symbiotic, representing a harmonious dance that influences the composition and function of the microbial community.

1. **Selective Nourishment:** Prebiotics act as selective nourishment for specific microbes, promoting the growth and activity of beneficial bacteria such as Bifidobacteria and Lactobacilli. This selective stimulation contributes to a balanced microbial community, preventing the overgrowth of potentially harmful bacteria.

2. **Fermentation and SCFA Production:** Upon reaching the colon, prebiotics undergo fermentation by gut bacteria, yielding SCFAs as metabolic byproducts. SCFAs, particularly acetate, propionate, and butyrate, play multifaceted roles in supporting gut health. They serve as energy sources for colonocytes, exert anti-inflammatory effects, and maintain a slightly acidic gut environment that favors beneficial microbes.

3. **Modulation of Microbial Composition:** Regular consumption of prebiotics has been linked to positive shifts in microbial composition. Enriching beneficial bacteria contributes to a diverse and resilient

microbiota, which is crucial for various aspects of health, including immune function and nutrient absorption.

Health Implications of Prebiotics

Prebiotics' impact extends beyond gut health, influencing various physiological systems and contributing to overall well-being.

1. Immune System Support: A robust and balanced gut microbiota is intricately linked to immune function. By promoting beneficial bacteria growth, prebiotics contribute to maintaining a healthy immune system. This support ranges from enhancing the activity of immune cells to influencing the production of immunomodulatory molecules.

2. Metabolic Health and Weight Management: Prebiotics affect metabolic health, impacting glucose metabolism and insulin sensitivity. The fermentation of prebiotics produces SCFAs that influence metabolic pathways, potentially contributing to weight management and the prevention of metabolic disorders.

3. Mood and Mental Well-Being: The gut microbiota influences the gut-brain axis, a bidirectional communication system between the gut and the central nervous system. Prebiotics, through their impact on microbial composition and the production of neurotransmitter precursors, may support mental well-being and mood regulation.

4. Bone Health: Emerging research suggests a connection between gut health and bone health, specifically the gut microbiota. By influencing mineral absorption and modulating inflammatory responses, prebiotics may contribute to maintaining healthy bones, especially in aging.

Practical Strategies for Incorporating Prebiotics

Integrating prebiotics into daily dietary practices requires a thoughtful approach considering individual preferences, dietary restrictions, and health goals.

1. **Diverse Dietary Sources:** A diverse array of foods contains prebiotics, allowing individuals to choose based on taste preferences and nutritional needs. A mix of fruits, vegetables, whole grains, and legumes provides a broad spectrum of prebiotic compounds.

2. **Embracing Fermented Foods:** In addition to directly consuming prebiotics, incorporating

fermented foods into the diet introduces live beneficial bacteria. Yogurt, kefir, sauerkraut, and kimchi not only provide prebiotics but also deliver probiotics, fostering a comprehensive approach to gut health.

3. **Prebiotic Supplements:** Prebiotic supplements offer a convenient option for individuals with specific health concerns or dietary restrictions. However, it is crucial to approach supplementation cautiously, considering individual needs and potential gastrointestinal sensitivities.

While prebiotics offer many health benefits, certain considerations and challenges merit attention.

1. Gastrointestinal Sensitivities: Individuals with certain gastrointestinal conditions, such as irritable bowel syndrome (IBS) or inflammatory bowel disease (IBD), may experience sensitivities to specific prebiotics. Tailoring prebiotic intake to individual tolerances and preferences becomes essential in such cases.

2. Balancing Prebiotics and FODMAPs: Certain prebiotics, such as fructans and galacto-oligosaccharides, belong to the category of fermentable oligosaccharides, disaccharides, monosaccharides, and polyols (FODMAPs).

Individuals sensitive to FODMAPs may need to balance obtaining prebiotics and managing symptoms, often requiring guidance from healthcare professionals.

3. Gradual Introductions: For those new to prebiotic-rich foods or supplements, a gradual introduction allows the gut to adapt to increased fibre intake. This approach minimizes the risk of gastrointestinal discomfort and supports the gradual establishment of a resilient microbial community.

Precision Prebiotics and Personalized Gut Health

As the field of microbiome research advances, future frontiers in prebiotics may witness the emergence of precision approaches tailored to individual microbial profiles.

1. **Microbiome Testing:** Advancements in microbiome testing technologies enable individuals to gain insights into their unique microbial composition. Precision prebiotics could be designed based on these profiles, targeting specific microbial needs for personalized gut health.

2. **Synergistic Combinations:** Research exploring the synergistic effects of combining specific prebiotics with probiotics or other bioactive compounds holds promise.

Understanding the nuanced interactions within the gut ecosystem may pave the way for precisely tailored interventions.

Chapter Four

Gut Health and Chronic Diseases: Breaking the Link

In the intricate tapestry of human health, the gut emerges as a central player, orchestrating a symphony of functions that extend far beyond digestion. This chapter will discuss the profound interconnection between gut health and chronic diseases, unraveling how maintaining a healthy gut microbiota can serve as a potent preventive measure against age-related illnesses. From cardiovascular diseases to neurodegenerative conditions, the gut is key to breaking the link with chronic diseases, offering a holistic approach to health and vitality.

Unraveling the Profound Interconnection Between Gut Health and Chronic Diseases

The human body is an intricately woven tapestry of interconnected systems, and at the epicenter of this biological symphony lies the gut – a bustling ecosystem teeming with trillions of microorganisms collectively known as the gut microbiota. Recent scientific endeavors have shed light on the profound interconnection between gut health and chronic disease development, progression, and prevention. This chapter delves deeper into the intricate web of interactions, exploring how the gut microbiota is pivotal in the genesis and mitigation of chronic conditions that often accompany aging.

Before delving into the interplay between gut health and chronic diseases, it's crucial to understand the dynamic nature of the gut microbiota. Comprising bacteria, viruses, fungi, and other microorganisms, this thriving community forms a symbiotic relationship with the human host. The gut microbiota is fundamental in various physiological processes, including nutrient metabolism, immune modulation, and maintaining a balanced inflammatory response.

Inflammation: A Double-Edged Sword

Inflammation, often described as the body's response to injury or infection, is a central theme in the interconnection between gut health and chronic diseases. While acute inflammation is a protective mechanism, chronic, low-grade inflammation can be detrimental and is recognized as a common denominator in many chronic conditions.

The gut microbiota plays a pivotal role in modulating the delicate balance between pro-inflammatory and anti-inflammatory responses. Dysbiosis, an imbalance in the composition of the gut microbiota, can lead to chronic inflammation, providing fertile ground for the initiation and progression of various diseases.

Cardiovascular Diseases: Beyond Cholesterol

Cardiovascular diseases, encompassing conditions such as atherosclerosis, heart failure, and stroke, have long been associated with factors like high cholesterol and hypertension. However, emerging research indicates that the gut microbiota contributes significantly to cardiovascular health.

The gut microbiota influences the metabolism of dietary compounds, producing metabolites that impact cholesterol levels, blood pressure, and arterial health. Moreover, the microbiota's role in regulating inflammation has far-reaching implications for cardiovascular diseases, as chronic inflammation is a key driver in developing atherosclerotic plaques.

Type 2 Diabetes: A Microbial Influence on Metabolism

Type 2 diabetes, characterized by insulin resistance and elevated blood sugar levels, is intricately linked to the gut microbiota's role in metabolism. The microbiota influences the breakdown of complex carbohydrates, impacting the production of short-chain fatty acids (SCFAs) that, in turn, influence glucose metabolism.

Moreover, the gut microbiota communicates with fatty tissue, modulating the release of hormones that play a role in insulin sensitivity. Dysbiosis has been associated with impaired glucose tolerance, emphasizing the pivotal role of gut health in mitigating the risk of type 2 diabetes.

Neurodegenerative Diseases: The Gut-Brain Axis in Focus

The gut-brain axis, a bidirectional communication system between the gut and the central nervous system, has gained prominence in understanding the pathogenesis of neurodegenerative diseases such as Alzheimer's and Parkinson's.

The gut microbiota plays a role in producing neurotransmitters and neuroactive compounds that influence brain function. Additionally, the microbiota regulates the permeability of the gut lining, influencing the entry of inflammatory molecules into the bloodstream. Chronic neuroinflammation, often observed in neurodegenerative diseases, can be influenced by the gut microbiota, opening avenues for preventive interventions by promoting gut health.

Cancer: Microbial Influences on Carcinogenesis

Cancer, characterized by uncontrolled cell growth, is a multifaceted condition influenced by genetic, environmental, and lifestyle factors. Recent research has spotlighted the role of the gut microbiota in modulating the risk of certain cancers.

The microbiota impacts inflammation, immune responses, and the metabolism of dietary compounds that may have carcinogenic or protective effects. Understanding and manipulating the gut microbiota may offer novel strategies for cancer prevention and adjuvant therapy.

Inflammatory Bowel Diseases (IBD): A Feedback Loop of Dysbiosis

Inflammatory bowel diseases, including Crohn's disease and ulcerative colitis, exemplify the bidirectional relationship between gut health and chronic illnesses. While these conditions are characterized by chronic inflammation in the gastrointestinal tract, dysbiosis often precedes and perpetuates the inflammatory cascade.

Disrupted gut microbiota can compromise the integrity of the gut lining, allowing the entry of harmful substances and perpetuating inflammation. Conversely, chronic inflammation in IBD can further alter the composition of the gut microbiota, creating a feedback loop that sustains the disease process.

Aging and Gut Health: Navigating the Complex Terrain

As individuals age, the gut microbiota undergoes changes influenced by many factors, including diet, medications, and lifestyle. The aging process is associated with alterations in immune function and physiological resilience, impacting the delicate balance of the gut ecosystem.

Recognizing the importance of gut health in ageing is paramount for promoting longevity and resilience against age-related chronic diseases. Nutritional considerations, physical activity, and a holistic approach to maintaining a diverse gut microbiota become essential components of healthy aging.

The Promise of Precision Medicine in Gut Health

As the field of microbiome research advances, precision medicine holds the promise of tailoring interventions based on individual microbial profiles. Microbiome testing technologies enable individuals to gain insights into their unique gut microbiota composition, paving the way for personalized strategies to promote gut health.

Emerging therapies, including precision probiotics and microbial therapies such as fecal microbiota transplantation (FMT), offer targeted approaches to manipulating gut microbiota for therapeutic benefits. These advancements signify a paradigm shift in preventive healthcare, where precision interventions aim to address the root causes of chronic diseases.

Strategies for Promoting Gut Health

Fostering a healthy gut microbiota takes center stage in the quest for optimal well-being. The symbiotic relationship between the gut and its microbial inhabitants has far-reaching implications for overall health. Here, we explore a comprehensive set of strategies, constituting a holistic approach, to promote and maintain a resilient and diverse gut microbiota.

1. **Dietary Interventions: A Nutrient-Rich Symphony**

   The cornerstone of gut health lies in dietary choices. A diet rich in diverse nutrients fuels the human host and its microbial companions. The following dietary principles contribute to a flourishing gut microbiota:

   Fiber-Rich Foods: Whole grains, fruits, vegetables, legumes, and nuts provide a buffet of fibers that nourish beneficial bacteria. These fibers undergo fermentation in the colon, producing short-chain fatty acids (SCFAs) supporting gut health.

   Prebiotic-Rich Choices: Prebiotics are non-digestible compounds that selectively nourish beneficial microbes. Foods such as garlic, onions, leeks, asparagus, and bananas contain prebiotics, fostering a thriving microbial community.

Probiotic-Rich Foods: Incorporating fermented foods introduces live beneficial bacteria into the gut. Yogurt, kefir, sauerkraut, kimchi, and miso are examples of probiotic-rich choices that contribute to microbial diversity.

Diverse Nutrient Intake: A varied diet ensures that the body receives a spectrum of nutrients, supporting overall health and microbial diversity. A rainbow of fruits and vegetables provides an array of vitamins, minerals, and antioxidants.

2. **Probiotics and Prebiotics: Synergistic Support**

Supplementation with probiotics and prebiotics offers targeted support for the gut microbiota. These interventions enhance the microbial balance, promoting the growth of beneficial bacteria.

Probiotic Supplements: These supplements contain live beneficial bacteria, such as Lactobacillus and Bifidobacterium strains. Probiotics can be particularly beneficial after a course of antibiotics or to address specific gut health concerns.

Prebiotic Supplements: For those seeking to augment their prebiotic intake, supplements containing non-digestible fibres like inulin or fructooligosaccharides offer a convenient option. These compounds selectively nourish beneficial microbes.

3. Regular Exercise: Movement for Microbial Diversity

Physical activity is vital for cardiovascular health and weight management and influences the gut microbiota. Regular exercise has been associated with increased microbial diversity and a more balanced microbial community.

Aerobic and Resistance Training: Aerobic exercises, such as running or cycling, and resistance training, like weightlifting, contribute to a healthy gut microbiota. Incorporating a mix of these activities supports overall well-being.

Consistency Matters: Regularity in physical activity has a positive impact on the gut microbiota. Establishing a routine that includes a variety of exercises enhances microbial diversity over time.

4. Stress Management: Balancing the Gut-Brain Axis

Chronic stress can disrupt the delicate interplay between the gut and the brain, influencing the gut microbiota and contributing to gastrointestinal issues. Managing stress is integral to maintaining a harmonious gut environment.

Mindfulness and Meditation: Practices such as mindfulness, meditation, and deep-breathing exercises have been linked to improvements in gut health. These techniques help modulate the stress response and promote a balanced gut-brain axis.

Regular Relaxation: Incorporating relaxation techniques into daily life, whether through hobbies, nature walks, or other enjoyable activities, contributes to stress reduction and supports gut health.

5. **Adequate Sleep: Nourishing the Microbial Garden**

Quality sleep is essential for overall health, and its impact extends to the gut microbiota. Sleep influences circadian rhythms and the production of hormones that, in turn, affect the microbial community.

Consistent Sleep Schedule: Maintaining a regular sleep-wake cycle supports the body's internal clock and contributes to a balanced gut microbiota. Aim for a consistent sleep schedule, even on weekends.

Create a Restful Environment: A conducive sleep environment, including a comfortable mattress and a dark, quiet room, enhances sleep quality. Adequate rest nourishes both the body and the microbial garden within.

6. **Limiting Antibiotic Use: Prudent Prescriptions**

Antibiotics, while crucial for combating bacterial infections, can inadvertently disrupt the balance of the gut microbiota. Prudent and mindful use of antibiotics helps minimize unintended consequences on gut health.

Follow Prescribed Guidelines: Adopt the recommended dosage and duration when prescribed antibiotics. Completing the full course helps eradicate the targeted bacteria and minimizes potential disruptions to the gut microbiota.

Probiotics During Antibiotic Use: In certain situations, healthcare providers may recommend probiotic supplementation alongside antibiotics to help mitigate the impact on the gut microbiota.

7. Hydration: Water for Wellness

Adequate hydration is fundamental for overall health and supports a conducive environment for the gut microbiota. Water helps transport nutrients, facilitates digestion, and maintains mucosal integrity.

Balanced Fluid Intake: Consuming appropriate water daily ensures optimal hydration. Herbal teas and infused water with fruits or herbs can add variety while contributing to hydration.

8. Diversified Food Sources: A Culinary Symphony

Introducing diverse foods into the diet provides an array of nutrients and compounds that benefit the gut microbiota. Experimenting with different cuisines and incorporating various ingredients enriches the microbial tapestry.

Global Cuisine Exploration: Trying foods from different cultures introduces new flavors and nutrient profiles. Each cuisine brings unique components that can contribute to a diverse gut microbiota.

Seasonal and Local Produce: Embracing seasonal and locally sourced produce ensures a fresh and varied intake of fruits and vegetables. Local markets often offer a spectrum of options that change with the seasons.

9. Nutritional Considerations for Seniors: Tailoring the Approach

As individuals age, nutritional needs may evolve, and considerations for gut health become even more crucial. Tailoring dietary choices to meet seniors' specific needs supports maintaining a healthy gut microbiota.

Protein-Rich Foods: Seniors may benefit from an increased intake of protein-rich foods to support muscle health. Sources like lean meats, dairy, legumes, and nuts contribute to a balanced diet.

Calcium and Vitamin D: Supporting bone health is paramount for seniors, and adequate calcium and vitamin D intake are essential. Dairy products, leafy greens, and fortified foods contribute to these nutrients.

10. Environmental Exposures: Mindful Living

The environment can impact the gut microbiota, and adopting mindful practices in daily living contributes to overall well-being. Considerations such as avoiding excessive use of disinfectants and embracing nature support a balanced microbial environment.

Green Spaces and Nature Exposure: Spending time in natural settings has been associated with positive effects on the gut microbiota. Whether a stroll in a park or gardening, connecting with nature benefits mental and microbial health.

Moderate Disinfectant Use: While hygiene is crucial, excessive use of disinfectants may negatively impact microbial diversity. Adopting a balanced approach to cleanliness helps maintain a healthy microbial balance.

In conclusion, the profound interconnection between gut health and chronic diseases unravels a narrative that extends far beyond conventional paradigms of disease prevention. With its intricate influence on inflammation, metabolism, and immune responses, the gut microbiota emerges as a central player in shaping the landscape of chronic diseases.

Chapter Five

The Gut-Brain Connection: Enhancing Cognitive Function

In the intricate dance between the gut and the brain, the profound impact of gut health on mental clarity and cognitive function emerges as a cornerstone of overall well-being. In this chapter, we will delve more into the intricacies of this relationship, shedding light on how the gut, often referred to as the "second brain," plays a pivotal role in shaping cognitive outcomes and mental acuity. Let's embark on a journey to unravel the profound impact a healthy gut can have on mental clarity and cognitive function.

Microbial Architects of Cognitive Harmony

Neurotransmitters and Microbial Symphony

At the heart of the gut-brain connection lies the synthesis of neurotransmitters—a symphony orchestrated by the microbial architects residing in the gut. These microscopic inhabitants actively produce neurotransmitters like serotonin, dopamine, and gamma-aminobutyric acid

(GABA), influencing mood, motivation, and cognitive function.

Serotonin, often associated with mood regulation, is not solely produced in the brain; a significant portion originates in the gut. The intricate interplay between the gut microbiota and the synthesis of neurotransmitters becomes a dynamic force in shaping mental clarity and emotional well-being.

Impact on Cognitive Resilience

Cognitive function, encompassing processes such as memory, attention, and problem-solving, is intricately tied to gut health. The gut microbiota's influence on inflammation, oxidative stress, and the production of neuroactive compounds creates an environment that either supports or hinders cognitive resilience.

In conditions of dysbiosis, where there is an imbalance in the gut microbiota, disruptions to cognitive function can occur. Conversely, a diverse and balanced microbial community contributes to an environment conducive to mental clarity, enhanced memory retention, and optimal cognitive performance.

Inflammation and Cognitive Impacts

Chronic inflammation, a common feature of an imbalanced gut microbiota, can profoundly affect cognitive function. Inflammation contributes to the breakdown of the blood-brain barrier, allowing inflammatory molecules to enter the brain and trigger neuroinflammation—a process implicated in neurodegenerative conditions.

The gut microbiota produces anti-inflammatory compounds like short-chain fatty acids (SCFAs) and is a frontline defender against excessive inflammation. A balanced gut microbiota actively contributes to maintaining optimal inflammatory responses, providing a shield against cognitive decline.

The Gut-Brain Axis in Aging

As individuals age, the gut microbiota changes, and these alterations are intimately linked to cognitive aging. The symbiotic relationship between the gut and the brain becomes a focal point for understanding the nuances of cognitive decline and the potential avenues for intervention.

In aging, a delicate interplay marks the gut-brain axis. Proactive measures to support a healthy gut, such as dietary interventions and lifestyle adjustments, emerge as crucial strategies for mitigating cognitive decline and preserving mental clarity in the golden years.

Dietary Strategies for Cognitive Wellness

Crafting a diet that nurtures the gut and the brain becomes a fundamental strategy for promoting cognitive wellness. Nutrient-dense foods that support the gut microbiota and provide essential brain nutrients take centre stage in this dietary symphony.

1. **Omega-3 Fatty Acids:** In fatty fish, flaxseeds, and walnuts, omega-3 fatty acids contribute to brain health and may have anti-inflammatory effects in the gut. These essential fats become building blocks for cognitive resilience.

2. **Antioxidant-Rich Foods:** Berries, leafy greens, and colourful vegetables are rich in antioxidants that combat oxidative stress. These compounds protect neurons from damage, contributing to cognitive well-being.

3. **Fiber and Whole Grains:** Nourishing the gut microbiota with a diverse array of fibers from whole grains and fiber-rich foods supports microbial

diversity. The fermentation of these fibers produces SCFAs, which play a role in maintaining cognitive function.

Probiotics and Prebiotics: Allies for Cognitive Health

Supplementation with probiotics and prebiotics offers targeted support for the gut microbiota, enhancing the microbial symphony that influences cognitive outcomes.

- **Probiotics:** Live beneficial bacteria, such as Lactobacillus and Bifidobacterium strains, contribute to cognitive health. Probiotic supplements and fermented foods provide additional sources for introducing these beneficial microbes into the gut.
- **Prebiotics:** Non-digestible fibres like inulin and fructooligosaccharides act as prebiotics, selectively nourishing beneficial microbes. These compounds serve as the fuel for the microbial orchestra, contributing to a harmonious gut-brain symphony.

Lifestyle Factors: Orchestrating Cognitive Well-Being

While dietary choices are pivotal, lifestyle factors contribute to the holistic orchestration of cognitive well-being through the gut-brain axis.

- **Physical Activity:** Regular exercise enhances microbial diversity and promotes blood flow to the brain, supporting cognitive function. The rhythmic cadence of physical activity becomes a melody for the gut and the brain.

- **Quality Sleep:** Adequate and restful sleep is fundamental for overall health, and its impact extends to the gut-brain connection. Sleep influences circadian rhythms, hormonal balance, and the consolidation of memories, contributing to cognitive well-being.

- **Stress Management**: Chronic stress can disrupt the delicate balance of the gut-brain axis. Practices such as meditation, mindfulness, and deep-breathing exercises contribute to stress reduction, fostering a harmonious connection between the gut and the brain.

Personalized Strategies for Mental Clarity

In pursuing mental clarity and cognitive resilience, personalized strategies take precedence. Recognizing the individuality of gut microbiota profiles, health status, and dietary preferences becomes paramount in crafting interventions that resonate with each person's unique needs.

Precision interventions, guided by insights into the individual gut microbiota, offer a pathway to tailor strategies for cognitive well-being. From personalized dietary plans to targeted probiotic interventions, these approaches acknowledge the diverse landscape of microbial communities and aim to optimize their potential for mental clarity.

Unveiling Transformative Potential

As we stand on the precipice of a new era in gut-brain research, the future holds the promise of transformative potential. Ongoing research endeavors unveil frontiers in understanding the intricate connections between gut health, mental clarity, and cognitive function.

Therapeutic possibilities, guided by a deepened understanding of the gut-brain axis, may open avenues for interventions that transcend traditional approaches. The prospect of targeted therapies, precision interventions, and novel insights into the microbial symphony within the gut offers a glimpse into the transformative journey ahead.

The Role of Microbial Anti-Inflammatories

Chronic inflammation, a hallmark of ageing, plays a central role in age-related cognitive decline. The gut microbiota, with its diverse community of microorganisms, becomes a key player in modulating inflammatory responses. Beneficial microbes actively produce anti-inflammatory compounds, such as short-chain fatty acids (SCFAs), traverse the gut-brain axis and exert neuroprotective effects.

These microbial anti-inflammatories serve as guardians, helping to quell the flames of neuroinflammation that can contribute to cognitive impairment. Nurturing a gut environment that fosters the production of SCFAs becomes a strategic approach to supporting cognitive resilience in the face of aging.

Microbial Influence on Blood-Brain Barrier Integrity

Preserving cognitive function requires addressing inflammation and maintaining the integrity of the blood-brain barrier. This critical barrier regulates the passage of substances between the bloodstream and the brain, safeguarding the delicate neural environment.

The gut microbiota contributes to the integrity of the blood-brain barrier through various mechanisms. Beneficial microbes enhance the production of molecules that support the barrier's structural integrity, preventing the entry of harmful substances into the brain. This microbial influence becomes a cornerstone in the symphony of cognitive resilience, fortifying the brain against potential insults.

Synthesizing Cognitive Harmony

Neurotransmitters, the messengers orchestrating communication between nerve cells, are pivotal in shaping cognitive function. With its ability to synthesize neurotransmitters, the gut microbiota becomes a key contributor to this cognitive symphony.

Serotonin, often called the "feel-good" neurotransmitter, is a product of microbial synthesis within the gut. Dopamine and GABA, crucial for motivation and emotional balance, also find their origins in the microbial community. The delicate balance of these neurotransmitters, influenced by the gut microbiota, becomes a dynamic force in nourishing the brain and supporting cognitive function.

Crafting a Brain-Friendly Plate

Dietary choices emerge as a central player in promoting cognitive well-being through the gut. Crafting a brain-friendly plate involves incorporating nutrient-dense foods that support gut microbiota and essential brain functions.

- **Omega-3 Fatty Acids:** In fatty fish, flaxseeds, and walnuts, omega-3 fatty acids contribute to brain health and may have anti-inflammatory effects in the gut. These essential fats become building blocks for cognitive resilience.

- **Antioxidant-Rich Foods:** Berries, leafy greens, and colorful vegetables abound in antioxidants that combat oxidative stress. By reducing the impact of free radicals, these foods contribute to overall well-being, including cognitive health.

- **Whole Grains and Fiber:** Nourishing the gut microbiota with a diverse array of fibers from whole grains and fiber-rich foods supports microbial diversity. This, in turn, influences the production of beneficial metabolites that contribute to cognitive resilience.

Supplementation with probiotics and prebiotics emerges as a targeted strategy to enhance the gut-brain connection and support cognitive health.

- **Probiotics:** Live beneficial bacteria, such as Lactobacillus and Bifidobacterium strains, offer potential cognitive benefits. Probiotic supplements and fermented foods provide additional sources for introducing these beneficial microbes into the gut.
- **Prebiotics:** Non-digestible fibres like inulin and fructooligosaccharides act as prebiotics, selectively nourishing beneficial microbes. These compounds serve as the fuel for the microbial orchestra, contributing to a harmonious gut-brain symphony.

Beyond Dietary Considerations

While dietary choices are pivotal, lifestyle factors contribute to the holistic symphony of cognitive resilience.

1. **Physical Activity:** Regular exercise enhances microbial diversity and promotes blood flow to the brain, supporting cognitive function. The rhythmic

cadence of physical activity becomes a melody for the gut and the brain.

2. **Quality Sleep:** Adequate and restful sleep is fundamental for overall health, and its impact extends to the gut-brain connection. Sleep influences circadian rhythms, hormonal balance, and the consolidation of memories, contributing to cognitive well-being.

3. **Stress Management:** Chronic stress can disrupt the delicate balance of the gut-brain axis. Practices such as meditation, mindfulness, and deep-breathing exercises contribute to stress reduction, fostering a harmonious connection between the gut and the brain.

Tailoring Strategies for the Golden Years

In the context of men over 60, the quest for cognitive resilience takes on added significance. Personalized strategies that consider individual health profiles, dietary preferences, and lifestyle factors become instrumental in crafting a symphony that resonates with the unique needs of this demographic.

Precision interventions, guided by insights into the individual gut microbiota, offer a pathway to tailor strategies for cognitive well-being. From personalized dietary plans to targeted probiotic interventions, these approaches acknowledge the diverse landscape of microbial communities and aim to optimize their potential for cognitive resilience.

Chapter Six

Digestive Harmony

As men journey into their golden years, the symphony of life takes on a different tune, as does the intricate dance of digestive health. We will delve into the unique challenges men over 60 may encounter in their digestive journey, exploring age-related changes, medication impacts, and the prevalence of chronic conditions. This chapter serves as a guide, offering insights into tailored solutions, personalized approaches, and emerging trends that shape this demographic's digestive wellness landscape.

Understanding the Digestive Landscape in Aging

Age-Related Changes in Digestion

The aging process brings about many changes in the digestive system, impacting its efficiency and resilience. Among these changes is a natural decline in the production of digestive enzymes. Enzymes that play a pivotal role in breaking down food may become less abundant, potentially affecting the absorption of essential nutrients.

Additionally, gut motility, the rhythmic contractions that move food through the digestive tract, may slow down. This can contribute to challenges such as constipation and a sense of bloating or fullness after meals.

Alterations in Gut Microbiota Composition

The composition of the gut microbiota, a diverse community of microorganisms residing in the gastrointestinal tract, undergoes shifts with age. The balance between beneficial and potentially harmful microbes may be disrupted, influencing digestive function and overall health.

Maintaining a healthy and diverse gut microbiota becomes crucial in mitigating age-related changes and promoting optimal digestive wellness. Probiotics, which introduce beneficial bacteria to the gut, emerge as valuable allies in supporting microbial balance.

Medication Impact on Digestive Health

Men often contend with a higher prevalence of chronic conditions, leading to the use of various medications. Polypharmacy, the simultaneous use of multiple medications, can have implications for digestive health.

Certain medications, such as pain relievers, antacids, and prescription drugs for chronic conditions, may contribute to digestive challenges. For instance, using proton pump inhibitors (PPIs) to manage acid reflux can alter stomach acid levels, potentially impacting nutrient absorption.

Medication-Induced Constipation and Diarrhea

Common side effects of medications, particularly those affecting the gastrointestinal system, include constipation or diarrhea. Opioid pain medications, anticholinergic drugs, and certain antidepressants are known for their potential to cause constipation, while others, such as antibiotics, may lead to diarrhea by disrupting the balance of gut bacteria.

Managing medication-induced digestive challenges involves a delicate balance between addressing the primary health concerns and mitigating side effects that impact digestive comfort.

Prevalence of Chronic Conditions and Secondary Digestive Effects

- **Cardiovascular Disease and Dietary Considerations**

  Men often face an increased risk of cardiovascular disease, and dietary considerations become paramount in managing both cardiovascular health and digestive wellness. Dietary modifications to support heart health, such as reducing saturated fats and sodium, may inadvertently impact digestive comfort.

Balancing the nutritional needs for cardiovascular health with digestive considerations requires personalized approaches. Working with healthcare providers and dietitians ensures that dietary plans address both aspects of overall well-being.

- **Diabetes and Digestive Impact**

The prevalence of diabetes tends to rise with age, and this chronic condition can have secondary effects on digestive health. Diabetes can impact the nerves that control digestive functions, potentially leading to issues such as gastroparesis (delayed stomach emptying) and an increased risk of gastrointestinal infections.

Managing diabetes and its digestive implications involves integrating dietary management, blood sugar control, and lifestyle adjustments.

- **Arthritis and Dietary Challenges**

  Arthritis, a condition characterized by inflammation of the joints, can pose challenges in dietary choices. Men with arthritis may struggle to prepare certain foods or navigate dietary recommendations that support joint health.

  Adapting dietary choices to accommodate arthritis-related challenges while promoting digestive well-being requires personalized strategies. Collaborating with healthcare providers and nutrition experts ensures that dietary plans are tailored to individual needs.

Tailored Solutions for Men Over 60

1. **Probiotics for Gut Microbiota Support:** Recognizing the age-related shifts in gut microbiota composition, incorporating probiotics becomes a targeted strateg. Probiotic supplements or probiotic-rich foods such as yogurt, kefir, and fermented vegetables offer a source of beneficial bacteria that support microbial balance.

2. **Dietary Modifications for Nutrient Absorption:** Men can benefit from dietary modifications that

enhance nutrient absorption to address the potential decline in digestive enzyme production. Choosing nutrient-dense foods, incorporating enzyme-rich options like papaya and pineapple, and optimizing meal composition contribute to effective digestion and nutrient utilization.

3. **Fiber Intake for Bowel Regularity:** Bowel regularity is a common concern for men in this age group. Increasing fiber intake from fruits, vegetables, whole grains, and legumes supports healthy bowel movements. Adequate hydration complements fiber intake, promoting soft stools and efficient transit through the digestive tract.

4. **Mindful Eating Practices:** The importance of mindful eating practices grows with age, emphasizing the need for men to savor and appreciate their meals. Chewing food thoroughly, eating relaxed, and being attuned to hunger and fullness cues contribute to effective digestion and overall dining satisfaction.

5. **Regular Physical Activity:** Regular physical activity is a cornerstone of digestive wellness for men over 60. Exercise supports gut motility, reduces the risk of constipation, and contributes to overall

well-being. Incorporating activities that align with individual abilities, such as walking or gentle exercises, fosters digestive resilience.

6. **Medication Management and Professional Guidance:** Communication with healthcare providers is essential for managing medications effectively and addressing potential digestive side effects. Collaborating with healthcare professionals, including gastroenterologists and dietitians, ensures a comprehensive digestive health approach that considers primary health concerns and medication impacts.

7. **Screening for Colon Health:** Given the increased risk of colorectal issues with age, regular screenings for conditions like colorectal cancer become imperative. Discussing with healthcare providers to determine the most appropriate screening schedule based on individual health history and risk factors promotes proactive colon health management.

Lifestyle Modifications for Holistic Well-Being

Incorporating lifestyle modifications that address overall well-being positively impacts digestive health. Stress management techniques, quality sleep, and other holistic approaches contribute to a harmonious gut environment. Men over 60 can benefit from activities that promote relaxation and emotional well-being.

Personalized Approaches and Holistic Wellness

1. **Individualized Dietary Plans:** Crafting individualized dietary plans considers personal preferences, nutritional needs, and digestive considerations. Working with a dietitian to tailor meal plans ensures that men over 60 receive the necessary nutrients while accommodating specific digestive sensitivities.

2. **Monitoring and Adjusting:** Regular monitoring of digestive health allows for prompt adjustments to dietary and lifestyle interventions. Keeping a food diary, noting digestive symptoms, and being attentive to changes in bowel habits contribute to proactive management.

3. **Holistic Wellness Programs:** Participating in holistic wellness programs integrating nutrition, physical activity, and mindfulness offers a comprehensive approach to digestive health. These

programs can be tailored to the unique needs and preferences fostering a holistic sense of well-being.

Advances in Gut Microbiome Research

Ongoing research in gut microbiome science continues to unveil insights into the role of the microbiota in health and disease. Future interventions may leverage personalized microbiome data to develop targeted strategies for digestive health in men.

Telehealth and Digital Solutions

Integrating telehealth and digital solutions provides convenient access to healthcare resources for men in this age group. Virtual consultations with healthcare providers, digital health platforms, and mobile applications offer avenues for ongoing support and guidance.

Tailored Nutraceuticals

Advancements in nutraceuticals, including supplements tailored to digestive health, may become more prevalent. Tailored formulations that address the specific needs of individuals could play a role in supporting digestive resilience.

The digestive challenges faced by men over 60 are nuanced and multifaceted, influenced by age-related changes, medication impacts, and the prevalence of chronic conditions. Navigating this complex landscape requires a holistic approach that addresses individual needs and promotes well-being.

By incorporating probiotics, making dietary modifications for optimal nutrient absorption, increasing fiber intake, embracing mindful eating practices, engaging in regular physical activity, managing medications effectively, and seeking professional guidance, men over 60 can navigate digestive challenges with resilience and proactive wellness.

As we delve into the future, ongoing research, digital solutions, and personalized interventions hold the promise of shaping the landscape of digestive health for men over 60. Chapter Seven will cast its gaze into emerging trends, potential breakthroughs, and the transformative landscape that lies ahead for digestive wellness in this demographic.

Chapter Seven

Mindful Nutrition: Eating with Intention

In the golden years, the importance of mindful nutrition became increasingly profound for men over 60. This chapter explores the tailored application of mindful eating principles to the unique needs and experiences of men in this demographic. As the symphony of life continues, cultivating conscious eating habits takes center stage, offering a pathway to enhanced well-being, optimal digestion, and a more profound connection with the nourishment that sustains the body and soul.

The Significance of Mindful Nutrition in the Golden Years

1. **Meeting Unique Nutritional Needs:** As men traverse the landscape of aging, their nutritional needs evolve. Mindful nutrition acknowledges and addresses these changes, recognizing the importance of a diet rich in essential nutrients that support bone health, cognitive function, cardiovascular wellness, and overall vitality.

2. **Digestive Resilience:** Mindful eating fosters digestive resilience, an aspect particularly relevant

for men over 60. With age, the digestive system may change, such as a slowdown in enzyme production and alterations in gut motility. Approaching meals with mindfulness can optimize the digestive process, promoting nutrient absorption and minimizing discomfort.

3. **Emotional Well-Being:** The golden years often bring a tapestry of life experiences, including transitions into retirement, changes in social dynamics, and reflections on a life well-lived. Mindful nutrition becomes a companion in these transitions, offering a tool to navigate emotional well-being through a conscious relationship with food.

Mindful Eating Principles

1. **Embracing Present Moment Awareness:** Incorporating present-moment awareness into meals is a foundational principle of mindful. This involves savoring each bite, paying attention to flavors and textures, and relishing the act of nourishing the body.

By embracing the present moment, men can derive greater satisfaction from their meals.

2. **Non-Judgmental Observation of Food Choices:** Mindful eating encourages a non-judgmental observation of food choices. This involves letting go of restrictive dieting mentalities and embracing a more intuitive approach. It's about choosing foods that align with nutritional needs and personal preferences without attaching labels of guilt or virtue.

3. **Connecting with Hunger and Fullness Cues:** As metabolic processes may shift with age, reconnecting with hunger and fullness cues becomes pivotal. Mindful eating invites you to listen to their bodies, distinguishing between physical hunger and emotional cues. This awareness supports optimal portion control and promotes a balanced relationship with food.

4. **Mindful Meal Planning:** Mindful meal planning integrates intentionality into choosing, preparing, and enjoying food. This may involve considering specific nutritional needs, incorporating various nutrient-dense foods, and engaging in the pleasurable aspects of culinary preparation.

5. **Cultivating Gratitude for Nourishment:** Gratitude is a key element of mindful nutrition, reminding individuals to appreciate the abundance of nourishment available. Cultivating gratitude extends beyond the physical act of eating to encompass a broader appreciation for the journey that brought them to this stage of life.

Practical Applications of Mindful Eating

1. **Mindful Eating Rituals:** Establishing mindful eating rituals can be particularly beneficial for men in this demographic. This may involve creating a serene eating environment, incorporating calming rituals before meals, and savoring the flavors slowly. Rituals add a layer of intentionality to the act of nourishment.

2. **Social Connection and Mindful Dining:** The social aspect of dining holds significant value for men over 60. Mindful eating encourages meaningful connections during meals, whether shared with family, friends, or within a community. The joy of

shared meals contributes to both emotional and nutritional well-being.

3. **Mindful Hydration:** Incorporating mindful hydration practices aligns with the principles of conscious nourishment. Men over 60 can benefit from being attentive to their hydration needs, choosing water as a primary beverage, and appreciating the role of adequate hydration in overall health.

4. **Mindful Snacking and Portion Control:** Mindful eating extends to snacking and portion control. Men over 60 can apply these principles by choosing nutrient-dense snacks, being attentive to portion sizes, and recognizing hunger and fullness signals. This approach supports balanced energy intake throughout the day.

Mindful Eating as a Holistic Approach to Well-Being

1. **Physical Well-Being:** Mindful nutrition contributes to physical well-being by promoting optimal nutrient intake, supporting digestive health, and fostering

overall vitality. For men over 60, this holistic approach aligns with the desire for sustained physical wellness and the maintenance of functional abilities.

2. **Cognitive Function:** The intricate relationship between nutrition and cognitive function became a focal point in the golden years. Mindful eating principles, emphasizing nutrient-dense choices and a balanced approach, support cognitive health. This connection reinforces the understanding that what nourishes the body also nurtures the mind.

3. **Emotional Resilience:** Navigating the emotional landscape of the golden years involves embracing changes and celebrating life's journey. Mindful eating is a tool for emotional resilience, providing a mindful response to stress, fostering a positive relationship with food, and enhancing overall emotional well-being.

Mindful Nutrition: A Transformative Journey

As we engage in mindful nutrition, the dining table becomes a canvas for reflections on a life well-lived. Mindful eating invites a pause, a moment to appreciate the richness of experience and acknowledge the body's journey through the years.

Mindful nutrition transcends the physical act of eating to become a practice of nurturing both the body and soul. As we embark on a transformative journey that aligns with their unique needs, honoring the wisdom gained over the years and embracing the potential for continued well-being.

The legacy of mindful nutrition extends beyond the individual, influencing the broader narrative of health and wellness. As we adopt and share these principles, they contribute to a culture of intentional nourishment, leaving a legacy of mindful living for future generations.

How mindful nutrition contributes to optimal gut health and well-being

Mindful nutrition positively influences gut health and overall well-being through a series of interconnected mechanisms, from the choices made during meals to the intricate workings of the gut microbiota. Here's a detailed exploration of how mindful nutrition contributes to optimal gut health and well-being:

Fostering Bacterial Diversity

- **Mindful Food Choices:** Mindful nutrition emphasizes a diverse and balanced diet, incorporating a variety of whole foods such as fruits, vegetables, whole grains, and lean proteins. This diversity in food choices is a buffet for the gut microbiota, promoting the growth of various beneficial bacteria.

- **Prebiotics and Fiber:** Prebiotics, found in certain foods like garlic, onions, and bananas, act as nourishment for beneficial bacteria in the gut. Additionally, dietary fiber from fruits and vegetables supports regular bowel movements and provides the necessary substrates for the growth of a diverse microbial community.

Supporting Digestive Resilience

- **Optimizing Nutrient Absorption:** Mindful nutrition involves choosing foods that support optimal nutrient absorption. This is crucial for individual, as age-related changes in digestion may impact nutrient uptake. Nutrient-dense choices enhance the efficiency of nutrient absorption, supporting overall health.

- **Balancing Macronutrients:** Mindful nutrition encourages a balanced intake of macronutrients, including carbohydrates, proteins, and healthy fats. This balance contributes to sustained energy levels, supports metabolic processes, and aids in maintaining digestive function.

Nurturing the Gut-Brain Axis

- **Mood and Cognitive Function:** The gut and brain communicate bidirectionally through the gut-brain axis. Mindful nutrition, rich in omega-3 fatty acids, antioxidants, and other brain-supportive nutrients, positively influences mood and cognitive function. These nutrients contribute to a healthy brain and may mitigate the risk of age-related cognitive decline.

- **Mindful Eating Practices:** Mindful eating involves being fully present during meals, savoring each bite, and appreciating the sensory experience. This mindful approach fosters a positive relationship with food, reducing stress and supporting mental well-being through the gut-brain axis.

Mitigating Disruptions from Medications and Aging

- **Awareness of Medication Impact:** Mindful nutrition involves awareness of the potential impact of medications on gut health. Certain drugs, commonly prescribed for older individuals, may alter the gut microbiota composition. By considering these effects, individuals can make informed choices to mitigate disruptions.

- **Promoting Microbial Resilience:** Mindful nutrition recognizes the resilience of the gut microbiota. While aging may bring natural changes to the microbial composition, a resilient gut microbiota adapts. Nutrient-dense, diverse foods support microbial resilience, maintaining a healthy balance despite age-related shifts.

Beyond Probiotics: Unlocking the Power of Prebiotics

- **Holistic Approach:** Chapter 3 explored the power of prebiotics, which are non-digestible fibers that fuel beneficial bacteria. Mindful nutrition embraces this holistic approach, understanding that gut health is not solely about adding beneficial bacteria (probiotics) but also providing the necessary nutrients for their growth (prebiotics).

- **Dietary Strategies:** Incorporating prebiotic-rich foods, such as garlic, onions, asparagus, and bananas, becomes a strategic part of mindful nutrition. These foods contribute to a nourishing environment for beneficial bacteria, promoting a flourishing gut microbiota.

Preventing Chronic Diseases

- **Anti-Inflammatory Choices:** Mindful nutrition often involves choosing anti-inflammatory foods, such as fatty fish, nuts, seeds, and colorful fruits and vegetables. Chronic low-grade inflammation is associated with various age-related diseases. By making anti-inflammatory choices, individuals can reduce the risk of cardiovascular disease, diabetes, and inflammatory disorders.

- **Balancing Blood Sugar:** Mindful nutrition includes paying attention to the glycemic index of foods, which can help stabilize blood sugar levels. This is particularly relevant, as balanced blood sugar contributes to overall metabolic health and reduces the risk of age-related diseases.

Holistic Well-Being

- **Mindful Eating Practices:** Mindful nutrition goes beyond the nutritional content of food. It encompasses the entire eating experience, from choosing and preparing food to eating itself. Mindful eating practices, such as enjoying meals in a calm environment and being fully present, contribute to a positive and mindful relationship with food.

- **Emotional Resilience:** The holistic approach of mindful nutrition includes emotional well-being. Nourishing the body with wholesome, nutrient-dense foods can positively impact mood and emotional resilience, fostering a balanced and positive outlook on life.

In conclusion, mindful nutrition weaves a tapestry encompassing diverse dietary choices, awareness of medication impacts, a focus on gut-brain axis health, and a commitment to preventing chronic diseases. This holistic approach contributes to the resilience of the gut microbiota, supports overall well-being, and becomes a guiding principle for individuals, especially those over 60, seeking to embrace a mindful and nourishing approach to nutrition.

Chapter Eight

Nutrigenomics

In the quest for optimal health and well-being, this chapter delves into the groundbreaking realm of nutrigenomics, an emerging field that explores the intricate interplay between genetics and nutrition. As we journey into the world of personalized nutrition for gut resilience, we uncover surprising facts about the profound impact of genetics on individual nutritional needs. This chapter seeks to unravel the genetic symphony that influences how our bodies respond to the foods we consume, paving the way for a new era of tailored dietary approaches.

Understanding Nutrigenomics

Nutrigenomics represents the convergence of two powerful domains: nutrition and genomics. At its core, it investigates how individual genetic variations influence responses to dietary components. By deciphering the unique genetic code embedded in each person, nutrigenomics aims to unlock personalized insights into the ideal Diet for optimal health and gut resilience.

Genes are not static entities but dynamic players in the orchestra of life. Nutrigenomics explores how dietary choices can modulate gene expression, turning certain genes on or off. This dynamic interaction forms the basis of personalized nutrition, where the symphony of genes responds to the specific nutrients an individual's Diet provides.

The Surprising Fact: Genetic Variability in Micronutrient Metabolism

One of the surprising revelations uncovered by nutrigenomics is the significant genetic variability in micronutrient metabolism. Individuals can harbor genetic variations that influence how their bodies absorb, transport, and utilize essential vitamins and minerals. This means that what works as a nutritional powerhouse for one person might have a different impact on another due to their unique genetic makeup.

Example: Vitamin D Metabolism

Take, for instance, vitamin D metabolism, a crucial player in bone health, immune function, and overall well-being. Some individuals may possess genetic variations that affect their ability to convert sunlight or dietary sources of vitamin D into its active form. This can result in varying levels of vitamin D effectiveness, even when consuming seemingly adequate amounts. Nutrigenomics sheds light on these genetic nuances, highlighting the need for personalized approaches to ensure optimal nutrient utilization.

The Gut Microbiome: A Genomic Player

As we explore the intricate world of nutrigenomics, it becomes clear that our genetic makeup extends beyond our human genome to encompass the genomes of trillions of microorganisms residing in our gut—the gut microbiota. These microbial genomes play a crucial role in how our bodies interact with food, influencing digestion, nutrient absorption, and overall gut health.

Nutrigenomics emphasizes that the same dietary component may elicit diverse responses in different individuals based on their gut microbiome composition. For example, the metabolism of certain fibers into short-chain fatty acids, essential for gut health, can vary based on the specific bacterial populations present. This personalized response underscores the need to consider human and microbial genomic dimensions in tailoring nutritional recommendations.

Unraveling the Genetic Symphony: Applications in Personalized Nutrition

1. Genetic Testing for Personalized Insights

Advancements in genomic technologies have paved the way for accessible genetic testing, allowing individuals to unravel their unique genetic symphony. Researchers and healthcare professionals can provide personalized insights into an individual's ideal dietary patterns by analyzing specific genetic markers related to nutrient metabolism.

2. Tailoring Macronutrient Ratios

Nutrigenomics enables the customization of macronutrient ratios based on genetic factors. For example, individuals with certain genetic variations may be predisposed to metabolize fats more efficiently than vice versa. Tailoring macronutrient intake aligns with genetic predispositions, optimizing energy metabolism and supporting overall gut resilience.

3. Micronutrient Optimization

Understanding genetic variations in micronutrient metabolism empowers individuals to optimize their intake of vitamins and minerals. Personalized recommendations can guide individuals toward foods rich in specific nutrients that align with their genetic profile, ensuring efficient absorption and utilization.

4. **Adapting to Genetic Predispositions**

Genetic insights can shed light on predispositions toward certain health conditions or sensitivities. For instance, individuals with genetic variants impacting lactose metabolism may benefit from personalized strategies to manage dairy intake. Nutrigenomics offers a roadmap to adapt dietary choices in alignment with genetic predispositions, fostering gut resilience and overall well-being.

Addressing Challenges in Nutrigenomics

While nutrigenomics holds immense promise, it has challenges. Genetic information is complex, and the interactions between genes and nutrition are multifaceted. Interpretation of genetic data requires careful consideration of environmental factors, lifestyle choices, and the dynamic nature of gene expression.

Navigating these challenges is essential for the responsible application of nutrigenomics principles. Integrating genetics into personalized nutrition raises ethical considerations concerning privacy and informed consent. Safeguarding genetic information is paramount, and individuals must clearly understand how their data will be used. Ethical frameworks and guidelines ensure that nutrigenomics applications prioritize the well-being and autonomy of individuals.

Case Studies: Real-Life Applications of Nutrigenomics

This chapter explores real-life case studies showcasing the practical applications of nutrigenomics in precision nutrition. From athletes fine-tuning their diets for optimal performance to individuals managing chronic conditions through personalized dietary interventions, these cases illuminate the transformative potential of aligning nutrition with genetic insights.

Case 1: Athlete Performance Optimization

In sports, where every edge matters, nutrigenomics is pivotal in optimizing athlete performance. Genetic markers related to energy metabolism, nutrient utilization, and recovery are analyzed to tailor dietary plans that enhance endurance, reduce inflammation, and support overall athletic resilience.

Case 2: Managing Chronic Conditions

Individuals facing chronic conditions such as diabetes, cardiovascular disorders, or autoimmune diseases benefit from nutrigenomics-guided interventions. Genetic insights inform dietary strategies that consider unique metabolic pathways, sensitivities, and inflammation markers, contributing to more effective management and improved quality of life.

Integrating Nutrigenomics into Public Health

The potential of nutrigenomics extends beyond personalized nutrition to public health initiatives. By understanding population-level genetic variations, policymakers, and healthcare professionals can develop targeted interventions to address specific nutritional needs and enhance gut resilience on a broader scale.

Educational Initiatives for Nutrigenomic Literacy

As nutrigenomics becomes increasingly integrated into healthcare, education is crucial in fostering literacy and understanding. Public awareness campaigns, educational programs, and accessible resources contribute to an informed society capable of making empowered choices about nutrition and well-being.

The surprising fact that genetics significantly influences individual nutritional needs underscores the importance of moving beyond one-size-fits-all dietary recommendations. As we unravel the genetic symphony, personalized nutrition is a powerful tool for enhancing gut resilience and overall well-being.

In the dynamic interplay between genes, nutrition, and gut microbiota, nutrigenomics opens new frontiers for precision nutrition. From tailoring macronutrient ratios to optimizing micronutrient intake, the applications of nutrigenomics are vast and diverse. Real-life case studies illuminate how individuals, from athletes to those managing chronic conditions, can benefit from personalized dietary approaches informed by their unique genetic makeup.

As we embrace the future of nutrigenomics, ethical considerations, and educational initiatives become integral to its responsible integration into healthcare. By unlocking the potential of personalized nutrition guided by genetic insights, we embark on a journey toward a healthier, more resilient gut and a future where well-being is truly individualized.

Nutrigenomics and the Gut Microbiome

- **The Microbial Dimension**

A crucial aspect of nutrigenomics is its consideration of the gut microbiome, the diverse community of microorganisms in the gastrointestinal tract. The gut microbiome, with its genomic content, adds a layer of complexity to the nutrigenomic puzzle.

- **Bidirectional Communication**

The relationship between nutrigenomics and the gut microbiome is bidirectional. On one hand, our genetic makeup influences the composition and function of the gut microbiota. On the other hand, the activities of the gut microbiota can modulate gene expression and impact how our bodies respond to nutrients.

Implications for Gut Health

One of the primary implications of nutrigenomics for gut health is the ability to provide tailored dietary recommendations based on an individual's genetic profile. Understanding how genes influence nutrient metabolism allows for the customization of diets to optimize gut health.

Nutrigenomics enables the identification of genetic factors that may influence digestive processes, such as enzyme production, gut motility, and nutrient absorption. By addressing these genetic variations, personalized nutrition can enhance digestive resilience, reducing the likelihood of gastrointestinal issues.

Certain genetic markers identified through nutrigenomics may be associated with inflammation and immune responses in the gut. Personalized nutrition can target these markers, helping to modulate inflammation and support a balanced immune system within the gastrointestinal tract.

Both genetic and environmental factors influence the gut microbiome's composition and function. Nutrigenomics allows a more nuanced understanding of how an individual's genetic makeup shapes their microbial landscape. This insight can guide dietary interventions to promote microbial harmony associated with optimal gut health.

Nutrigenomics in Action
1. Precision Nutrition

Nutrigenomics provides the tools for precision nutrition, tailoring dietary recommendations to individual genetic profiles. This approach has practical applications in gut health, from addressing specific digestive challenges to promoting a balanced gut microbiome.

2. Managing Gastrointestinal Disorders

For individuals with gastrointestinal disorders, nutrigenomics can offer targeted strategies. Understanding how genetic factors contribute to conditions such as irritable bowel syndrome (IBS) or inflammatory bowel disease (IBD) allows personalized dietary interventions to manage symptoms and improve overall gut health.

3. Weight Management

Genetic variations influence how individuals respond to different macronutrients, impacting metabolism and satiety. Nutrigenomics can inform personalized weight management plans, considering an individual's genetic predispositions related to body composition and energy balance.

4. Gut Microbiome Modulation

Nutrigenomics shows how specific dietary components interact with the gut microbiome based on individual genetics. This knowledge can be leveraged to design diets that support the growth of beneficial microbes while mitigating the overgrowth of potentially harmful species.

Chapter Nine

Gut-Immune Resilience - Strengthening the Body's Defense

In the intricate dance of human health, the relationship between the gut and the immune system stands as a pivotal alliance. This chapter embarks on a comprehensive exploration of gut-immune resilience, delving into the interconnected pathways that fortify the body's defense mechanisms.

From the role of the gut microbiome in immune modulation to lifestyle factors influencing immune function, this chapter unveils the intricate tapestry that shapes our ability to ward off pathogens, prevent infections, and maintain overall well-being.

Understanding the Gut-Immune Connection

The gut, often referred to as the body's "second brain," is not only a hub for digestion but a central player in immune resilience. The vast surface area of the gastrointestinal tract is exposed to a myriad of microorganisms, necessitating a robust defense system.

The gut-immune connection goes beyond mere physical proximity; it represents an intricate communication network that orchestrates immune responses and maintains a delicate balance between defense and tolerance.

At the heart of gut-immune resilience lies the gut microbiome—a diverse community of microorganisms that populate the digestive tract. These microbial architects play a crucial role in shaping the immune landscape, influencing the development and function of immune cells, and fine-tuning responses to potential threats.

In the intricate dance of the gut-immune axis, the microbial symphony emerges as a captivating composition that orchestrates the delicate balance between defense and tolerance. The gut microbiome, a diverse ecosystem of trillions of microorganisms, takes center stage in this symphony, influencing immune modulation in profound ways.

Microbial Diversity: A Prelude to Immune Education

- **The Educational Ballet**

  Microbial diversity within the gut sets the stage for an educational ballet that shapes the immune system's understanding of self and non-self.

  The gut is not a sterile environment but a bustling ecosystem where various bacteria, viruses, fungi, and other microorganisms coexist.

  This rich diversity serves as a training ground for immune cells, teaching them to distinguish between harmless commensals and potential threats.

Tolerance Through Exposure

  Exposure to a broad spectrum of microbes promotes immune tolerance, especially early in life. Tolerant immune cells are less likely to mount aggressive responses to harmless substances, reducing the risk of allergies and autoimmune conditions.

  This process, known as immune education, is orchestrated by the microbial players in the gut, each contributing a unique note to the symphony.

Short-Chain Fatty Acids (SCFAs): Harmonizing Immune Responses

In the microbial symphony, metabolic byproducts take center stage, and one of the key orchestrators is short-chain fatty acids (SCFAs). These molecules are produced through the fermentation of dietary fiber by gut bacteria. As the metabolic overture begins, SCFAs emerge as powerful signaling molecules influencing immune responses.

Imagine SCFAs as conductors guiding immune cells to harmonize their responses. These molecules have a multifaceted impact on immune modulation:

- **Anti-Inflammatory Baton:** SCFAs wield an anti-inflammatory baton, directing immune cells to adopt anti-inflammatory profiles. This helps maintain a balanced immune response, preventing excessive inflammation that could lead to chronic conditions.

- **Regulatory T Cell Choreography:** SCFAs play a key role in orchestrating the dance of regulatory T cells. These cells are instrumental in suppressing excessive immune reactions and promoting immune tolerance, ensuring the immune system responds appropriately to threats while avoiding unnecessary attacks on the body's tissues.

- **Epigenetic Symphony:** SCFAs influence the epigenetic landscape of immune cells. By modifying gene expression patterns, SCFAs contribute to the programming of immune cells, fostering a state of readiness that allows for swift and appropriate responses to infections.

Gut Microbiome and Mucosal Immunity

The gut is lined with a specialized immune system known as mucosal immunity. This frontline defense mechanism operates in the mucosal lining of the gastrointestinal tract, where most microbial interactions occur. The gut microbiome and mucosal immunity engage in a harmonious partnership, working together to ensure effective defense without triggering unnecessary alarms.

Immunoglobulin A (IgA) takes center stage in the mucosal immunity symphony. Produced by immune cells in the gut, IgA is like a musical sonata, orchestrating a targeted defense against potentially harmful microbes. The gut microbiome influences the production of IgA, creating a protective barrier that prevents pathogens from breaching the mucosal lining.

While the microbial symphony is designed for harmony, disruptions can occur, leading to a state known as dysbiosis.

Dysbiosis represents a discordant note in the symphony, where the balance of beneficial and harmful microbes is disrupted. This imbalance can profoundly affect immune modulation, potentially contributing to inflammatory conditions and increasing infection susceptibility.

Dysbiosis may lead to immune dysregulation, where the finely tuned immune responses orchestrated by the gut microbiome are thrown into disarray. Inflammatory pathways may be overactivated, and the delicate balance of immune tolerance may be compromised. This dysregulation can contribute to the development or exacerbation of autoimmune conditions and chronic inflammatory diseases.

Age-Related Changes in the Microbial Symphony

As individuals age, the microbial symphony undergoes subtle but significant changes. These age-related shifts, often called microbial evolution, influence the composition and function of the gut microbiome. The symphony that once played a vibrant and diverse tune may experience a shift in key and tempo, requiring adaptation in immune modulation strategies.

Immunosenescence, the aging of the immune system, adds a layer of complexity to the microbial symphony. The conductor's baton changes hands, and the immune system may exhibit diminished responsiveness, making older individuals more susceptible to infections and less efficient in modulating immune responses.

Strategies to Harmonize the Microbial Symphony

1. **Dietary Crescendo**

   Diet becomes a crescendo in the microbial symphony. A diet rich in fiber, fruits, vegetables, and fermented foods contributes to microbial diversity and the production of SCFAs. This dietary crescendo supports immune modulation, fostering a balanced and resilient immune system.

2. **Probiotic and Prebiotic Duets**

   Probiotics and prebiotics can be likened to duets in the microbial symphony. Probiotics introduce beneficial strains of bacteria, enhancing microbial diversity and contributing to immune education. Prebiotics, on the other hand, provide the nourishment necessary for the growth of these beneficial microbes, sustaining the harmonious melody of the gut microbiome.

3. **Lifestyle Harmony**

Beyond Diet, lifestyle factors play crucial roles in maintaining the microbial symphony. Adequate sleep, stress management, and regular physical activity contribute to immune resilience. Sleep rejuvenates the body's defenses, stress management prevents dissonance, and exercise supports the circulation of immune cells, ensuring the symphony plays on in harmony.

Nutrition as the Cornerstone of Immune Resilience

Nutrition serves as the foundational building block for a robust immune system. A well-balanced and nutrient-dense diet provides the essential vitamins, minerals, and antioxidants necessary for the proper functioning of immune cells. These micronutrients act as cofactors in various immune processes, influencing everything from the production of antibodies to the maturation of immune cells.

Certain vitamins and minerals emerge as key players in immune health. Vitamin C, found in citrus fruits and bell peppers, supports immune cell function and is an antioxidant. Vitamin D, synthesized through sunlight exposure and found in fatty fish, modulates immune responses. Zinc, present in foods like nuts and legumes, is essential for immune cell proliferation and function.

The colorful array of fruits and vegetables brings a spectrum of antioxidants, each with its unique ability to neutralize free radicals. By mitigating oxidative stress, antioxidants contribute to a balanced immune response, preventing chronic inflammation that can compromise immune function.

Building on the insights from Chapter Three, probiotics and prebiotics take center stage in promoting immune resilience. Probiotics in fermented foods like yogurt and kimchi contribute to a balanced gut microbiome, enhancing immune modulation. Prebiotics, abundant in foods like garlic, onions, and bananas, provide the necessary fuel for beneficial microbes, fostering a harmonious environment for immune cells.

Sleep: The Immune Rejuvenator

Quality sleep emerges as a non-negotiable pillar of immune health. The circadian rhythm, the body's internal clock, regulates various physiological processes, including immune function. Disruptions to this rhythm, often seen in irregular sleep patterns or insufficient sleep duration, can compromise immune responses and increase susceptibility to infections.

Sleep Cycles and Immune Activities

During sleep, the body undergoes crucial processes that contribute to immune rejuvenation. The stages of sleep, particularly deep sleep (slow-wave sleep) and REM sleep, play roles in producing and releasing immune-signaling molecules. These molecules, including cytokines and antibodies, contribute to immune surveillance, pathogen recognition, and coordinating immune responses.

Chronic Sleep Deprivation: A Compromised Defense

Chronic sleep deprivation, prevalent in modern lifestyles, can have detrimental effects on immune function. Studies have shown that individuals who consistently experience inadequate sleep are more susceptible to infections, and their immune responses to vaccines may be less effective. Prioritizing sufficient and restorative sleep becomes a cornerstone in fortifying the body's defense against pathogens.

Physical Activity: Mobilizing Immune Defenses

Regular physical activity emerges as a potent modulator of immune function. Exercise contributes to immune surveillance by promoting the circulation of immune cells throughout the body. It also enhances the activity of natural killer cells, a crucial component of the innate immune system responsible for identifying and eliminating infected cells.

Exercise induces various immunomodulatory effects that contribute to immune resilience. It can reduce chronic inflammation linked to age-related diseases and immune dysfunction. Additionally, exercise enhances the production of anti-inflammatory cytokines, contributing to a balanced immune response.

As individuals age, the benefits of exercise on immune health become particularly relevant. Regular physical activity can counteract the effects of immunosenescence, the age-related decline in immune function. Tailored exercise routines, considering individual fitness levels and health conditions, become essential for promoting immune vigilance in older individuals.

Stress Management: Calming the Immune Storm

Chronic stress poses a significant challenge to immune vigilance. The release of stress hormones, such as cortisol, can suppress immune cell activity and alter the balance of immune responses. Prolonged stress can contribute to inflammation, creating an environment compromises immune function and increases susceptibility to infections.

Effective stress management strategies play a pivotal role in maintaining immune resilience. Mindfulness practices, such as meditation and deep breathing, have been shown to reduce stress levels and modulate immune responses. These techniques foster relaxation that positively influences the gut-immune axis and promotes overall well-being.

The Gut-Brain Axis: Stress as a Two-Way Street

The bidirectional communication between the gut and the brain, explored in Chapter Six, becomes particularly relevant in the context of stress and immune function. Stress signals from the brain can impact gut function, altering the composition of the gut microbiome and influencing immune responses. Conversely, gut-derived signals can influence stress responses, highlighting the interconnected nature of the gut-brain-immune axis.

Holistic Approaches to Lifestyle for Immune Resilience
Integration of Lifestyle Factors

A holistic approach to lifestyle for immune resilience involves integrating multiple factors. Rather than viewing nutrition, sleep, physical activity, and stress management in isolation, recognizing their synergistic effects creates a comprehensive strategy to fortify the body's defense.

As discussed in Chapter Eight on nutrigenomics, personalized approaches to lifestyle factors take into account individual genetic variations. Tailoring nutritional strategies, sleep recommendations, and exercise routines based on genetic insights enhances interventions' precision, optimizing each individual's immune function.

Immunosenescence: Navigating Immune Aging

Aging changes the immune system, a phenomenon known as immunosenescence. The gut-immune axis undergoes shifts, influencing the composition and function of immune cells in the gut mucosa. Understanding these changes becomes crucial for promoting immune resilience in the aging population.

Thymic Involution: Impact on Immune Vigilance

The thymus, a key organ for the maturation of T cells, undergoes involution with age, affecting the production of naive T cells. Strategies to support thymic function and boost the availability of diverse T cells contribute to enhanced immune vigilance in older individuals.

Beyond Pathogens: The Immune System in Autoimmunity and Chronic Diseases

Autoimmunity: A Dysregulated Defense

While the immune system's primary role is to protect against external threats, dysregulation can lead to autoimmune conditions. The gut-immune axis plays a crucial role in autoimmune diseases, with imbalances in the gut microbiome potentially contributing to the breakdown of immune tolerance.

Chronic Inflammation: The Silent Culprit

Chronic inflammation, a common feature of aging, is intricately linked to the gut-immune axis. The persistent activation of immune cells in response to low-grade stimuli can contribute to various chronic diseases, from cardiovascular conditions to neurodegenerative disorders.

Nutritional Strategies for Gut-Immune Resilience

Probiotics: Allies in Immune Defense

Chapter Three highlighted the power of probiotics, and their role in gut-immune resilience takes center stage. Probiotic supplements or fermented foods rich in beneficial bacteria contribute to immune modulation, promoting a balanced immune response and reducing the risk of infections.

Prebiotics: Nourishing the Immune Garden

Just as prebiotics foster bacterial diversity, they also nourish the immune system. Prebiotic-rich foods, such as garlic, onions, and bananas, provide the fibers necessary for the growth of beneficial microbes that, in turn, contribute to immune education and tolerance.

Antioxidants: Guardians Against Inflammation

Antioxidants in colorful fruits and vegetables serve as guardians against inflammation. By neutralizing free radicals and mitigating oxidative stress, antioxidants support a balanced immune response and reduce the risk of chronic inflammatory conditions.

The Role of Stress in Gut-Immune Dynamics

Chronic stress can exert profound effects on the gut-immune axis. The release of stress hormones, such as cortisol, can impact the composition of the gut microbiome and alter immune cell function. Stress management strategies, including mindfulness and relaxation techniques, become essential for maintaining gut-immune resilience.

Chapter Ten

Seasonal Eating - Aligning Diet with Nature

In the intricate dance between nature and nutrition, seasonal eating emerges as a harmonious rhythm that resonates with the cycles of the Earth. This chapter will look into the profound benefits of aligning dietary choices with seasonal changes, exploring how this age-old practice contributes to optimal health, sustainable ecosystems, and a connection to the natural world. From the nutritional richness of seasonal produce to the environmental impact of local sourcing, this chapter unravels the layers of wisdom embedded in the art of seasonal eating.

Seasonal eating is a dietary approach that involves choosing and consuming foods in season during a specific time of the year. It is based on the idea that certain fruits, vegetables, and other food items are naturally abundant and at their peak in terms of flavor, nutrition, and freshness during specific seasons.

Seasonal eating has deep roots in traditional agricultural practices and is influenced by plant growth and harvest cycles. Before the advent of globalized food systems, communities relied on locally available produce, which varied depending on the time of year and climate. Seasonal eating is a return to this practice, emphasizing consuming naturally available foods harvested at optimal times.

Key aspects of seasonal eating include:

- **Freshness and Nutrient Density:** Seasonal produce is often harvested at its peak ripeness, likely to be more flavorful and nutritionally dense. The fruits and vegetables have had time to fully develop their nutrients, providing a range of vitamins, minerals, and antioxidants.

- **Connection to Nature:** Seasonal eating encourages a connection to the natural world and its cycles. By aligning dietary choices with the changing seasons, individuals can appreciate the rhythm of nature and gain a deeper understanding of where their food comes from.

- **Environmental Sustainability:** Choosing locally sourced, seasonal foods can contribute to environmental sustainability. It reduces the carbon footprint associated with transportation, supports local farmers, and promotes biodiversity by celebrating a variety of crops that can thrive in different seasons.

- **Balanced Nutrition:** Seasonal eating inherently encourages a balanced and varied diet. Different seasons offer different types of produce, allowing individuals to rotate their food choices and benefit from a diverse range of nutrients throughout the year.

- **Culinary Creativity:** Embracing seasonal produce can inspire creativity in the kitchen. Experimenting with recipes based on what is available locally and in-season adds variety to meals and encourages an appreciation for the unique qualities of different foods.

- **Health and Well-Being:** Seasonal eating aligns with the body's natural needs at different times of the year. For example, consuming foods rich in vitamin C during the colder months can support the immune system. This alignment with nature's cycles is believed to contribute to overall health and well-being.

While seasonal eating has historical and cultural roots, it has gained renewed interest in recent years as part of a broader movement towards sustainable and mindful food choices. Many individuals and communities are exploring ways to incorporate seasonal eating into their lifestyles, whether by shopping at farmers' markets, participating in community-supported agriculture (CSA) programs, or growing their seasonal produce.

The Essence of Seasonal Eating

1. **A Time-Honored Tradition**

   Seasonal eating is a practice deeply rooted in the traditions of various cultures across the globe. Before the era of globalized food systems, communities relied on abundant local produce available during specific times of the year. This practice ensured a diverse and nutrient-rich diet and fostered a sustainable relationship with the environment.

2. **Nature's Bounty: A Symphony of Flavors**

   Each season brings forth a unique array of fruits, vegetables, and other food items. The diversity of flavors, textures, and colors during different times of the year provides a sensory experience beyond mere sustenance. Seasonal eating invites individuals to savor the distinct characteristics of each ingredient, celebrating the cyclical nature of agricultural abundance.

Nutritional Richness of Seasonal Produce

1. Freshness and Nutrient Density

One of the primary advantages of seasonal eating lies in the freshness and nutrient density of seasonal produce. Fruits and vegetables harvested in their natural seasons are likely to be at their peak ripeness, containing higher levels of vitamins, minerals, and antioxidants. Consuming these nutrient-dense foods supports overall health and well-being.

2. Adaptation to Seasonal Needs

Nature has a way of providing what the body needs at different times of the year. For example, citrus fruits like oranges and grapefruits are abundant in winter, providing a vitamin C boost during the cold and flu season. Harvested in the fall, root vegetables offer grounding and nourishment as the weather transitions to colder temperatures.

3. Balancing Nutrient Intake

Seasonal eating inherently encourages a balanced nutrient intake. In the summer, a bounty of hydrating fruits like watermelon and berries supports the body's need for increased hydration. In contrast, the heartier vegetables and grains available in the fall and winter provide the sustenance needed for colder weather.

Environmental Impact of Seasonal Eating

1. Local Sourcing: Reducing Food Miles

Embracing seasonal eating often goes hand in hand with supporting local agriculture. Choosing locally sourced produce reduces the carbon footprint associated with transportation, as the food can reach consumers easily. This conscious choice contributes to lower greenhouse gas emissions and promotes environmental sustainability.

2. Biodiversity and Ecosystem Health

Seasonal eating supports biodiversity by celebrating the natural diversity of crops that can thrive in different seasons. This diversity is essential for the health of ecosystems, as it helps prevent monoculture—the cultivation of a single crop over large areas—which can lead to soil depletion and increased vulnerability to pests.

3. Reduced Dependence on Artificial Inputs

When food is grown in its natural season and environment, there is less reliance on artificial inputs such as pesticides and synthetic fertilizers. Seasonal crops are better adapted to local conditions, reducing the need for chemical interventions. This, in turn, promotes soil health and preserves the delicate balance of local ecosystems.

Seasonal Eating and Personal Well-Being

1. Connection to Nature's Rhythms

Seasonal eating fosters a deeper connection to the natural world and its cyclical patterns. By aligning dietary choices with the seasons, individuals become attuned to the ebb and flow of nature, cultivating a sense of mindfulness and gratitude for the gifts each season brings. This connection to nature's rhythms has profound implications for mental and emotional well-being.

2. Harmony with Circadian Rhythms

Our bodies have internal circadian rhythms influencing various physiological processes, including digestion. Seasonal eating aligns with these circadian rhythms, promoting optimal digestion and assimilation of nutrients. This alignment supports overall metabolic health and can contribute to a more balanced and sustainable approach to weight management.

3. Immune Support through Seasonal Variability

Seasonal changes often coincide with fluctuations in immune challenges. Consuming various seasonal foods exposes the body to diverse nutrients, supporting the immune system's adaptability to seasonal threats. For example, incorporating immune-boosting foods during the colder months can help fortify the body against common winter illnesses.

Practical Tips for Embracing Seasonal Eating

1. Familiarize Yourself with Seasonal Cycles

Understanding the natural cycles of local produce helps individuals make informed choices about what to include in their diets at different times of the year. Seasonal calendars from local farmers' markets or agricultural extension services can serve as valuable guides.

2. Explore Farmers' Markets and Local Suppliers

Farmers' markets and local suppliers often showcase the best of each season's harvest. Exploring these options allows individuals to access fresh, locally-grown produce while supporting regional agriculture. Building relationships with local farmers can also provide insights into the growing practices and conditions of the food.

3. **Embrace Preservation Techniques**

Preserving seasonal produce through methods such as canning, freezing, or fermenting enables individuals to enjoy the nutritional benefits of seasonal foods beyond their peak harvest times. These preserved items can serve as vibrant additions to meals during seasons when certain fresh produce may be less abundant.

4. **Experiment with Seasonal Recipes**

Seasonal eating invites culinary creativity as individuals explore recipes that showcase the flavors of each season. Experimenting with seasonal ingredients not only adds variety to meals but also provides an opportunity to appreciate the unique qualities of different foods throughout the year.

Overcoming Challenges in Globalized Food Systems

In a world where globalization has enabled access to a wide variety of foods year-round, embracing seasonal eating may present challenges. However, individuals can strive to strike a balance by prioritizing local, seasonal options when available and making conscious choices about imported foods.

Chapter Eleven

Fasting for Gut Renewal

The concept of intermittent fasting has emerged as a powerful tool for gut renewal. This chapter delves into the intricacies of intermittent fasting, exploring its profound impact on gut health, microbiota diversity, and overall well-being. From the science behind fasting-induced autophagy to the potential therapeutic applications for digestive disorders, this chapter unravels the transformative potential of embracing periods of abstention for gut renewal.

Intermittent fasting is a dietary approach that alternates between periods of eating and fasting. Unlike traditional diets that focus on specific foods or caloric intake, intermittent fasting is more about when you eat. It doesn't prescribe specific foods but designates specific time windows for eating and fasting.

There are several popular methods of intermittent fasting, each with its unique approach to structuring eating and fasting periods. Here are some common types:

Time-Restricted Eating (TRE):

- **Method:** Limiting daily food intake to a specific time window, typically 8 to 12 hours.
- **Example:** A common approach is the 16/8 method, where you fast for 16 hours and eat during an 8-hour window.

Alternate-Day Fasting:

- **Method:** Alternates between days of regular eating and days of significant calorie restriction or complete fasting.
- **Example:** On fasting days, individuals may consume very few calories or abstain from food altogether.

5:2 Diet:

- **Method:** Involves eating normally for five days of the week and significantly restricting calorie intake (usually around 500-600 calories) on two non-consecutive days.
- **Example:** Eat normally on Monday, Wednesday, Friday, Saturday, and Sunday, and restrict calories on Tuesday and Thursday.

Periodic Prolonged Fasting:

- **Method:** Includes extended fasting periods, often 24 to 72 hours or even longer. This approach is typically done less frequently, such as once a month or every few months.

- **Example:** Fasting for an entire day or more on specific occasions.

It's important to note that during fasting periods, individuals are generally advised to stay hydrated by drinking water, herbal teas, or other non-caloric beverages. Fasting does not mean complete abstinence from fluids.

Key Principles of Intermittent Fasting

1. **No Caloric Restriction:** Intermittent fasting doesn't necessarily involve restricting the types of foods consumed or counting calories. Instead, it focuses on when to eat.

2. **Flexibility:** There is flexibility in choosing the fasting and eating windows based on individual preferences and lifestyle. The goal is to find a sustainable pattern that works for the individual.

3. **Adaptation:** It may take some time for the body to adapt to intermittent fasting. Individuals may initially experience hunger during fasting, but many find these sensations diminish over time.

4. **Health Monitoring:** Individuals with underlying health conditions or concerns should consult with healthcare professionals before starting intermittent fasting. Monitoring health markers and overall well-being is important, especially for those with specific medical conditions.

Benefits of Intermittent Fasting

1. **Weight Management:** Intermittent fasting can be effective for weight loss by reducing overall calorie intake and promoting fat utilization during fasting periods.

2. **Improved Insulin Sensitivity:** Fasting periods may improve insulin sensitivity, helping regulate blood sugar levels and reducing the risk of type 2 diabetes.

3. **Cellular Repair and Autophagy:** Fasting induces autophagy, a cellular process that removes damaged or dysfunctional cellular components, contributing to cellular repair and renewal.

4. **Gut Health:** Intermittent fasting has been associated with positive effects on gut health, including microbiota diversity and gut barrier integrity improvements.

5. **Heart Health:** Some studies suggest that intermittent fasting may have cardiovascular benefits, including improved blood lipid profiles and reduced risk factors for heart disease.

6. **Brain Health:** Fasting may have cognitive benefits, including improved brain function and protection against age-related neurodegenerative diseases.

It's essential to approach intermittent fasting with a balanced and individualized perspective. What works for one person may not work for another, and the sustainability of the chosen fasting pattern is key for long-term success. Consulting with healthcare professionals, especially for individuals with existing health conditions, is advisable before embarking on an intermittent fasting regimen.

Benefits of intermittent fasting for gut health

Intermittent fasting has gained popularity for its potential benefits in weight management and its positive impact on various aspects of health, including gut health. This dietary approach involves cycling between periods of eating and fasting, and research suggests several benefits for the gut microbiota and overall digestive wellness.

1. Microbiota Diversity:

a. Promotion of Beneficial Bacteria:

Intermittent fasting has been associated with an increased abundance of beneficial bacteria in the gut, such as Bifidobacteria and Lactobacillus. These microbes play a crucial role in maintaining gut health by contributing to dietary fiber fermentation and producing short-chain fatty acids (SCFAs).

b. Microbial Resilience:

Fasting periods are a mild stressor for the gut microbiota, promoting microbial resilience. The temporary nutrient deprivation challenges the microbial community, encouraging the survival and proliferation of species better adapted to fluctuating conditions. This enhanced resilience contributes to a more robust and adaptable gut microbiome.

2. Gut Barrier Integrity:

a. Renewal and Repair:

Intermittent fasting has been linked to improvements in gut barrier integrity. The fasting-induced stress triggers cellular renewal and repair processes within the gut lining. This renewal helps to strengthen the integrity of the gut barrier, reducing the likelihood of inflammation and leaky gut syndrome.

b. Tight Junction Maintenance:

Tight junctions are protein structures that form a barrier between intestinal cells, controlling the passage of molecules. Intermittent fasting may contribute to maintaining tight junction integrity, preventing the unwanted passage of toxins and harmful substances through the gut barrier.

3. Inflammation Reduction:

a. Anti-Inflammatory Effects:

Chronic inflammation is linked to various digestive disorders. Intermittent fasting has demonstrated anti-inflammatory effects, potentially reducing inflammation in the gut. This anti-inflammatory action contributes to a healthier gut environment and may benefit individuals with inflammatory bowel disease (IBD) conditions.

b. Cytokine Modulation:

Fasting influences the production and release of cytokines, which are signaling molecules involved in immune responses. Modulating cytokine levels through intermittent fasting may contribute to a balanced immune response in the gut, helping to manage inflammation and support overall gut health.

4. Regulation of Gut Hormones:

a. Ghrelin and Leptin Balance:

Ghrelin and leptin are hormones that play roles in hunger and satiety. Intermittent fasting can influence the balance of these hormones, improving sensitivity to hunger and fullness cues. This hormonal regulation may contribute to better control of food intake and support overall digestive wellness.

b. Insulin Sensitivity:

Fasting periods lead to decreased insulin levels, promoting insulin sensitivity. Improved insulin sensitivity is associated with better glucose regulation and metabolism. This can positively impact conditions like metabolic syndrome, which is often linked to digestive health issues.

5. **Autophagy:**

a. Cellular Recycling:

Autophagy, a cellular recycling process, is upregulated during fasting. This process involves the removal of damaged or dysfunctional cellular components. In gut health, autophagy contributes to the renewal of intestinal cells, supporting the maintenance of a healthy gut lining.

b. Tissue Repair:

Fasting-induced autophagy extends to tissues and organs, including the gut. The removal of cellular waste and the recycling of components contribute to tissue repair and regeneration. This process supports the overall health and functionality of the digestive system.

6. **Metabolic Flexibility:**

a. Shift to Fat Utilization:

During fasting, the body shifts from using glucose as its primary energy source to utilizing stored fat. This metabolic flexibility promotes the breakdown of fats, generating ketones that can be used for energy. This shift in metabolism may have positive implications for gut health, as it reduces reliance on constant glucose availability.

b. Mitochondrial Health:

Fasting has been associated with improvements in mitochondrial health. Mitochondria are the energy-producing structures within cells. Enhanced mitochondrial function supports the energy needs of cells in the gut, contributing to the overall vitality of digestive tissues.

Gut Renewal Through Intermittent Fasting

The gut microbiota, a diverse community of microorganisms in the digestive tract, plays a crucial role in gut health. Intermittent fasting has been shown to impact the composition and diversity of the gut microbiota positively. Fasting periods create a dynamic environment that supports the growth of beneficial microbes while reducing the abundance of potentially harmful species.

Intermittent fasting is a stressor for the gut microbiota, promoting microbial resilience. The temporary nutrient deprivation challenges the microbial community, encouraging the survival and proliferation of species better adapted to fluctuating conditions. This enhanced resilience contributes to a more robust and adaptable gut microbiome.

The gut barrier, consisting of a layer of cells and mucous membranes, acts as a protective barrier between the gut lumen and the rest of the body. Intermittent fasting has been associated with improvements in gut barrier integrity. The renewal and repair processes triggered during fasting contribute to a healthier, more resilient gut lining.

Therapeutic Potential for Digestive Disorders

- **Inflammatory Bowel Disease (IBD)**

  Intermittent fasting shows promise as a complementary approach for individuals with inflammatory bowel diseases (IBD), such as Crohn's disease and ulcerative colitis. The anti-inflammatory effects of fasting and improvements in gut microbiota diversity may contribute to symptom relief and support the management of chronic inflammatory conditions.

- **Irritable Bowel Syndrome (IBS)**

  For individuals with irritable bowel syndrome (IBS), characterized by symptoms like abdominal pain and altered bowel habits, intermittent fasting offers a potential avenue for symptom management. The regulated eating windows and periods of fasting help mitigate symptoms associated with gut hypersensitivity and dysregulated motility.

- **Gut-Brain Axis and Mental Health**

   The gut-brain axis, explored in previous chapters, is a bidirectional communication system between the gut and the central nervous system. Intermittent fasting may influence the gut-brain axis, potentially impacting mental health conditions such as anxiety and depression. Research in this area suggests a complex interplay between fasting-induced changes in the gut microbiota and neurotransmitter production.

Practical Approaches to Intermittent Fasting

1. Time-Restricted Eating

Time-restricted eating involves limiting the daily eating window to a specific time frame, such as 8 hours, and abstaining from food for the remaining 16 hours. This approach is one of the most commonly practiced forms of intermittent fasting and can be adapted to individual preferences and lifestyles.

2. **Alternate-Day Fasting**

Alternate-day fasting consists of alternating between days of regular eating and days of significant calorie restriction or complete fasting. Individuals may consume a very low-calorie diet on fasting days or abstain from food altogether. This approach provides more extended periods of fasting, allowing for enhanced autophagy and metabolic flexibility.

3. **Periodic Prolonged Fasting**

Periodic prolonged fasting involves more extended fasting periods, ranging from 24 to 72 hours or even longer. This approach is typically done less frequently, such as once a month or once every few months. Prolonged fasting provides a more profound autophagic response and may be explored for its potential therapeutic effects.

Integrating Intermittent Fasting into a Healthy Lifestyle

Intermittent fasting is most effective when integrated into a holistic approach to health. Combining fasting with other lifestyle practices, such as mindful nutrition, regular physical activity, and stress management, creates a comprehensive framework for overall well-being.

Sustainability is a key consideration when adopting intermittent fasting as a lifestyle. The fasting approach should align with individual preferences, daily routines, and long-term goals. Fostering a positive relationship with food and maintaining a healthy mindset is essential for intermittent fasting.

Chapter Thirteen

Gut Health and Physical Performance

Maintaining physical vitality becomes a cornerstone of well-being in aging, particularly for men over 60. In this chapter, we will discuss the connection between gut health and physical performance, unraveling the symbiotic relationship that influences the digestive system and the overall vitality of the body. From nutrient absorption to the impact of gut microbiota on energy metabolism, this chapter explores how a healthy gut contributes to sustaining physical vigor in the golden years.

Nutrient Absorption and Energy Production

The digestive system, particularly the gut, plays a central role in absorbing essential nutrients from our food. Vitamins, minerals, and amino acids are critical for energy production, muscle function, and overall physical performance.

Microbial Influence on Metabolism

The gut microbiota, a diverse community of microorganisms in the intestines, influences metabolism. Certain microbial species contribute to the breakdown of complex carbohydrates, production of short-chain fatty acids

(SCFAs), and regulation of energy metabolism—all factors that impact physical vitality.

Facets that highlight the significance of maintaining a healthy gut for optimal well-being

1. Nutrient Absorption and Muscle Health

a. Sarcopenia and Nutrient Utilization:

Sarcopenia, the age-related loss of muscle mass and strength, becomes a pertinent concern. A healthy gut is essential for optimal nutrient absorption, ensuring essential vitamins and minerals for muscle health are absorbed efficiently. The decline in stomach acid and enzyme production associated with aging can impact nutrient utilization, making it imperative to support digestive processes for sustained muscle function.

b. Protein Absorption and Synthesis:

Protein, a key macronutrient for muscle repair and synthesis, requires effective digestion and absorption. A well-functioning gut ensures that the amino acids derived from protein digestion are absorbed into the bloodstream, supporting the maintenance and repair of muscle tissue. Strategies such as incorporating high-quality protein sources and optimizing digestive efficiency become pivotal for combating age-related muscle loss.

2. Microbial Influence on Metabolism

a. Mitochondrial Health and Energy Production:

The gut microbiota contributes to mitochondrial health, the powerhouse of cells responsible for energy production. Supporting mitochondrial function becomes crucial for sustaining physical vitality. Certain microbial species play a role in the breakdown of dietary components, influencing energy metabolism and contributing to overall energy levels. A diverse and balanced gut microbiome may enhance the efficiency of energy production, promoting sustained physical performance.

b. Short-Chain Fatty Acids (SCFAs) and Energy Utilization:

SCFAs, produced by gut bacteria during the fermentation of dietary fibers, have been associated with improved energy utilization. In the aging process, where energy requirements may fluctuate, the role of SCFAs in supporting metabolic flexibility becomes particularly relevant. Enhancing the production of these beneficial compounds through dietary choices fosters an environment that positively influences energy metabolism.

3. Gut Microbiota and Exercise Response

a. Enhanced Endurance and Recovery:

The gut microbiota's influence on the body's response to exercise extends beyond youth into the golden years. Maintaining physical vitality involves optimizing exercise benefits. A diverse and balanced gut microbiome has been linked to improved endurance, faster recovery, and reduced inflammation. These factors contribute to enhanced exercise capacity and resilience, fostering a sustainable approach to physical activity.

b. Inflammatory Modulation for Joint Health:

Chronic inflammation can contribute to joint discomfort and impact physical mobility. The gut-body axis regulates inflammation, and a healthy gut microbiota can contribute to modulating inflammatory responses. Experiencing age-related joint concerns, supporting gut health is a potential avenue for addressing inflammatory factors affecting physical comfort and mobility.

4. **Mind-Gut Connection and Mental Resilience:**

a. Cognitive Function and Physical Performance:

The connection between the gut and the brain, often called the gut-brain axis, is a dynamic interplay that influences mental well-being and cognitive aspects of physical performance. Cognitive function, including focus, coordination, and reaction time, is integral to optimal physical vitality. A healthy gut supports the production of neurotransmitters and neuroactive compounds that contribute to mental resilience during physical activities.

b. Stress Response and Physical Well-Being:

Physical or psychological stress can impact gut health and vice versa. Chronic stress can contribute to digestive issues, disrupting the balance of the gut microbiota. Conversely, an imbalanced gut microbiota may influence stress responses. Navigating the complexities of aging, strategies to manage stress and support gut health become intertwined elements in promoting overall physical well-being.

5. Personalized Nutrition for Gut-Driven Vitality

a. Individual Variability in Gut Health:

Recognizing the individual variability in gut health is fundamental. You may experience diverse digestive profiles influenced by genetics, lifestyle, and health history. Embracing personalized nutrition tailored to individual gut health considerations ensures that dietary choices align with specific requirements for sustained physical performance.

b. Addressing Digestive Challenges:

Age-related changes in digestion, coupled with potential digestive challenges, necessitate a proactive approach to address specific needs. You may benefit from targeted dietary interventions, including incorporating easily digestible foods, prebiotic-rich sources, and probiotics to support digestive harmony and gut health.

Practical Strategies for Gut-Driven Vitality

1. **Nutrient-Rich Diet:** A diet rich in nutrient-dense foods, including fruits, vegetables, lean proteins, and whole grains, supports gut health and provides the essential building blocks for physical vitality. Nutrient variety ensures a broad spectrum of micronutrients contributing to optimal bodily function.

2. **Prebiotics for Microbial Well-Being:** Prebiotics, non-digestible fibers that serve as food for beneficial gut bacteria promote microbial well-being. Including prebiotic-rich foods, such as garlic, onions, and bananas, in the diet fosters an environment that supports the growth of beneficial microbes.

3. **Hydration and Digestive Harmony:** Adequate hydration is essential for digestive harmony. Water supports the transport of nutrients, the breakdown of food particles, and the overall efficiency of the digestive process. Ensuring proper hydration is a foundational element of sustaining physical vitality.

4. **Physical Activity and Microbiota Diversity:** Regular physical activity has been associated with increased microbial diversity in the gut. The reciprocal relationship between exercise and gut

health underscores the importance of a holistic approach that integrates both elements for optimal physical performance.

5. **Mindful Eating Practices:** Mindful eating involves paying attention to the sensory experience of eating, including flavors, textures, and chewing. This practice enhances the enjoyment of meals and supports the digestive process, contributing to nutrient absorption and overall gut health.

Gut Health Challenges in Aging

1. **Age-Related Changes in Digestion:**

Aging brings about physiological changes in the digestive system, including declining stomach acid production and reduced enzyme activity. These changes can affect nutrient absorption and digestion efficiency, necessitating dietary adjustments to support gut health.

2. **Medications and Digestive Impact:**

Certain medications commonly prescribed for age-related conditions may have implications for gut health. For instance, using proton pump inhibitors (PPIs) to manage acid reflux can alter the gut microbiota and potentially impact nutrient absorption. Awareness of medication-related effects on the gut is crucial for addressing digestive challenges.

Integrating Gut Health into Holistic Well-Being

Sustaining physical vitality requires a holistic approach that considers the interconnectedness of gut health with other aspects of well-being. Addressing stress management, sleep quality, and mental health contributes to a comprehensive strategy for promoting overall vitality.

Recognizing individual variations in gut health and nutritional needs is fundamental. Personalized nutrition, tailored to an individual's unique digestive profile and health status, ensures that dietary choices align with specific requirements for sustained physical performance.

Gut Health Monitoring and Optimization

Regular health checkups provide opportunities to assess gut health and address emerging concerns. Monitoring digestive symptoms, such as changes in bowel habits, bloating, or discomfort, helps identify potential issues that can be addressed proactively.

Functional testing, including assessments of gut microbiota composition and digestive function, can offer insights into the specific dynamics of gut health. These tests enable a more nuanced understanding of individual requirements for optimization.

Chapter Fourteen

Gut Health and Longevity Practices
Around the World

Exploring the profound relationship between gut health and longevity unveils a rich tapestry woven by diverse cultures around the globe. This chapter delves into the fascinating realm of traditional practices, dietary habits, and lifestyle choices integral to promoting gut health and fostering longevity across different regions.

From the Mediterranean to Okinawa, from fermented foods in Asia to probiotic-rich choices in Europe, this chapter embarks on a global journey to unravel the wisdom embedded in age-old practices that contribute to vibrant digestive wellness and extended lifespans.

Unveiling Timeless Wisdom: The Global Tapestry of Gut Health

As men embark on the remarkable aging journey, the quest for longevity becomes a shared aspiration across cultures. This chapter will unravel the rich tapestry of gut health and longevity practices, drawing inspiration from diverse traditions and wisdom cultivated worldwide. This chapter serves as a compass, navigating through time-honored practices encapsulating sustained well-being and vitality.

1. Mediterranean Mastery: Nourishing the Gut with Abundance

a. The Mediterranean Diet:

The shores of the Mediterranean have long been a beacon of healthful living. The Mediterranean diet, renowned for its emphasis on fresh fruits, vegetables, olive oil, and lean proteins, takes center stage. Adopting this diet nourishes the gut with fiber-rich foods and provides many antioxidants that combat oxidative stress, contributing to gut resilience and overall longevity.

b. Fermented Wonders: Yogurt and Beyond:

The Mediterranean region boasts a rich tradition of fermented foods, notably yogurt. Beyond its probiotic prowess, yogurt and other fermented delicacies form a cornerstone of gut health. The symbiotic dance of beneficial bacteria in fermented foods supports a balanced gut microbiome, offering men a time-tested ally in the pursuit of digestive vitality.

2. Asian Elixirs: Fermentation as a Fountain of Wellness

a. Kimchi and Miso Magic:

From the vibrant streets of Seoul to the serene landscapes of Kyoto, Asia reveals its fermented treasures. Kimchi, a spicy Korean delicacy, and miso, a staple in Japanese cuisine, exemplify the art of fermentation. Packed with probiotics, these fermented wonders fortify the gut microbiota, promoting diversity and resilience. You can glean from these age-old practices, incorporating fermented delights for a gut brimming with vitality.

b. Tea Time Traditions: Kombucha and Beyond:

The Far East unfolds its tea-time traditions, with kombucha taking center stage. Originating in China and revered for its probiotic content, kombucha epitomizes the marriage of culture and gut health. Rich in polyphenols and beneficial microbes, fermented teas offer a refreshing elixir that nurtures the gut and fosters a harmonious balance.

3. **Latin American Legends: Beans, Plantains, and Probiotics**

a. The Power of Beans:

Steeped in tradition, Latin American cuisine imparts timeless wisdom for gut health. Beans, a nutritional cornerstone, bring soluble fibers that nurture gut bacteria. As we often savor the flavors of black beans, pinto beans, and more, they harness the gut-nourishing potential embedded in the region's culinary traditions.

b. Plantains and Prebiotics:

In the heart of Latin America, plantains emerge as a nutritional powerhouse. Beyond their delectable taste, plantains offer prebiotic fibers that fuel the growth of beneficial gut bacteria. Exploring the role of plantains in Latin American culinary heritage, uncover a natural source of prebiotic goodness, supporting the intricate dance of microbes within.

4. Nordic Resilience: Omega-3s, Fermented Fish, and Rye Bread

a. Omega-3 Rich Delights:

With their icy landscapes, the Nordic regions give the world a lesson in resilience. Fatty fish, abundant in omega-3 fatty acids, are a testament to the Nordic approach to gut health. Incorporating omega-3-rich delights like salmon and mackerel becomes a strategy for nurturing gut integrity and reaping the anti-inflammatory benefits that transcend the ages.

b. Fermented Fish Fare: Surströmming and More:

The art of fermentation expresses itself in Nordic fish dishes like surströmming. As an acquired taste for some, these fermented delicacies contribute to the diversity of the gut microbiota. Exploring the Nordic culinary landscape, we often embark on a journey of discovery, embracing fermented fish as a conduit to gut resilience.

5. African Abundance: Plant Diversity and Gut Harmony

a. Okra, Millet, and Sorghum Symphony:

Across the vast expanse of Africa, a myriad of plant-based delights weaves a tapestry of gut health. Okra, a fiber-rich vegetable, promotes digestive regularity in this symphony. Millet and sorghum, in African diets, offer nutrients that fortify the gut lining. You would find inspiration in the diverse plant offerings of Africa, aligning with nature's bounty for sustained well-being.

b. Teff Triumph: Fermentation in Ethiopian Traditions:

With its rich cultural heritage, Ethiopia introduces teff, a tiny grain with a colossal impact on gut health. Injera, a fermented flatbread crafted from teff, epitomizes Ethiopian culinary wisdom. The fermentation process enhances nutrient bioavailability and introduces probiotics, contributing to the flourishing landscape of the gut microbiome.

6. **Oceanic Wisdom: Seaweed, Kelp, and the Bounty of the Sea**

   a. Seaweed Sensations:

Oceanic communities from the Pacific Islands to New Zealand's shores share their gut health wisdom. Seaweed, abundant in essential minerals and prebiotic fibers, emerges as a staple. They are exploring the benefits of seaweed tap into a natural prebiotic source that promotes the growth of beneficial gut bacteria, fostering a resilient digestive terrain.

b. Kelp Elegance: A Gift from the Depths:

With its majestic presence beneath the ocean waves, Kelp offers a glimpse into the secrets of gut health. Rich in fucoidans and alginates, kelp exhibits anti-inflammatory and prebiotic properties. As you can embrace the elegance of kelp in their diet, they draw from the ocean's bounty to support gut harmony and overall longevity.

7. Indigenous Insights: Root Vegetables, Berries, and Gut Resilience

a. Roots of Well-Being:

Indigenous communities around the world cultivate the wisdom of root vegetables. From sweet potatoes to yams, these nutrient-dense roots contribute fibers and antioxidants that foster gut resilience. Incorporating the roots of well-being becomes a tribute to the time-tested practices of indigenous cultures, fortifying the gut for enduring health.

b. Berry Brilliance: A Colorful Medley:

Berries, vibrant in antioxidants, punctuate the dietary traditions of indigenous populations. Whether it's the maqui berries of Chile or the açaí berries of the Amazon, these colorful fruits offer phytonutrients that support gut health. Embrace the brilliance of berries, infusing their diet with nature's vibrant palette for sustained well-being.

8. Ayurvedic Harmony: Spices, Herbs, and the Gut-Body Balance

a. Turmeric Triumph: The Golden Elixir:

Ayurveda, the ancient healing system of India, imparts timeless wisdom for gut health. At the heart of Ayurvedic traditions lies turmeric, a golden elixir revered for its anti-inflammatory and gut-soothing properties. Embracing turmeric unlock the potential of this spice to promote digestive harmony, aligning with the principles of Ayurveda for holistic well-being.

b. Digestive Elixirs: Cumin, Coriander, and Fennel:

The triumvirate of cumin, coriander, and fennel forms a trinity of digestive elixirs in Ayurvedic practices. These aromatic seeds, when combined, create a blend known as "CCF tea." Sipped throughout the day, CCF tea supports digestion, reduces bloating, and nurtures gut balance. Incorporate these digestive allies, attuning their bodies to the rhythmic wisdom of Ayurveda.

9. European Traditions: Fermented Cheeses, Sauerkraut, and Digestive Legacy

a. Fermented Cheese Charisma:

The rolling hills of Europe unfold a treasure trove of fermented cheeses, each bearing a distinctive character. From Gouda to Roquefort, fermented cheeses contribute not only to gastronomic delight but also to gut health. Savoring the richness of fermented cheeses partake in a centuries-old legacy that embraces the art of fermentation for digestive wellness.

b. Sauerkraut Symphony: A Culinary Classic:

Sauerkraut, a staple in European cuisines, encapsulates the alchemy of fermentation. Cabbage transformed by beneficial microbes into sauerkraut becomes a probiotic-rich condiment. As we explore the sauerkraut symphony, we tend to connect with the digestive legacy of European traditions, where fermented foods stand as testaments to the interplay between culture and gut health.

10. Global Emissaries: The Role of Travel, Adaptogens, and Mindful Practices

a. Travel as a Catalyst for Gut Diversity:

The act of travel transcends borders, offering a passport to gut diversity. Exposing the gut to new cuisines, local flavors, and microbial landscapes contributes to a more resilient gut microbiome. From street markets in Marrakech to ramen stalls in Tokyo, the global gastronomic journey becomes a canvas for nurturing gut health through culinary exploration.

b. Adaptogens for Stress Resilience:

Adaptogens, revered in traditional medicine systems globally, emerge as allies in the quest for gut and overall well-being. Herbs like ashwagandha, rhodiola, and holy basil adapt to the body's needs, supporting stress resilience. Navigating the complexities of life, adaptogens become integral components of a holistic approach, fostering gut harmony in the face of life's demands.

c. Mindful Practices: From Yoga to Tea Ceremonies:

Beyond the plate, mindful practices enrich the landscape of gut health. Yoga, with its emphasis on breath and movement, supports digestive function. Tea ceremonies, revered in cultures like Japan and China, offer a space for mindful consumption. Incorporating these practices into their routine cultivate a mindful approach to nourishment, nurturing the gut-body connection with intention and presence.

Embracing Culinary Diversity

1. **Fermentation Festival: A Global Culinary Carnival**

From the kimchi of Korea to the sauerkraut of Germany, fermentation emerges as a common thread weaving through culinary traditions globally. You can embark on a flavorful journey, indulging in the tangy allure of fermented foods that populate tables from Seoul to Strasbourg. The fermentation festival celebrates the microbial magic that transforms humble ingredients into probiotic-rich delights, fostering gut health and culinary adventure.

2. **Spice Bazaar: A Global Expedition of Flavor**

Spice markets, vibrant and aromatic, beckon men of senior years to explore the world through a palette of colors and scents. From the turmeric of India to the sumac of the Middle East, spices elevate dishes and contribute to digestive harmony. The spice bazaar celebrates the global symphony of flavors that enliven meals, enhancing both taste and gut well-being.

1. **Garden of Gut Wellness: A Harvest from Every Continent**

Cultivated with care and reverence, gardens offer a cornucopia of ingredients promoting gut health. Cultivating your gut wellness gardens can embrace a diversity of plants—from Brazil's açaí palms to the Mediterranean rosemary bushes. The garden becomes a microcosm of the world's botanical richness, nurturing not just plants but also the microbial life within the soil and, by extension, the gut.

2. **Farmers' Market Exploration: From Teff to Tomatoes**

Farmers' markets, bustling with life and color, invite you to meander through stalls adorned with global treasures. Teff from Ethiopia, tomatoes from Italy, and yuca from Brazil converge in a celebration of agricultural diversity. The market becomes a vibrant testament to the resilience of local farming practices worldwide, offering a bounty that supports gut health through a spectrum of nutrient-rich offerings.

Cultural Connections and Traditions

1. **Tea Ceremony: Sipping Wisdom from Asia to Europe**

   The ritual of tea transcends continents, embodying not just a beverage but a philosophy of mindfulness. Participating in a global tea ceremony—from Japan's matcha rituals to England's afternoon teas—immerse themselves in a practice beyond hydration. The tea ceremony celebrates various herbal infusions' presence, connection, and potential gut-friendly properties.

2. **Yoga Retreat: A Journey of Mind-Body Harmony**

   Yoga, originating in ancient India, has become a global practice fostering holistic well-being. Joining yoga retreats can experience the unity of mind and body, recognizing the gut-brain axis as a vital component of overall health. The yoga retreat becomes a celebration of movement, breath, and the interconnectedness of bodily systems.

Mindful Integration of Traditional Wisdom

1. **Ancient Wisdom in Modern Kitchens: Ayurvedic Cooking Classes**

Ayurvedic cooking classes offer you an opportunity to weave ancient Indian wisdom into modern kitchens. From balancing doshas to incorporating gut-friendly spices, these classes become a celebration of culinary alchemy guided by Ayurvedic principles. The kitchen transforms into a laboratory where flavors and gut health harmonize, embracing a holistic approach to nourishment.

2. **Traditional Healing Modalities: From Acupuncture to Ayahuasca**

Beyond culinary practices, traditional healing modalities worldwide contribute to celebrating gut health. Acupuncture from China, Ayahuasca ceremonies from South America, and African indigenous herbal remedies represent a kaleidoscope of approaches. Exploring these modalities connect with diverse perspectives on holistic health, recognizing the intricate connections between the gut and broader well-being.

Community Celebrations
1. **Gut Health Potluck: A Feast of Global Delights**

Hosting a gut health potluck becomes a communal celebration of diverse practices. Friends and neighbors bring dishes inspired by their cultural backgrounds—fermented kimchi, Greek yogurt, Indian dal—all contributing to a vibrant spread. When you are surrounded by this global feast, you foster a sense of community while embracing the shared goal of gut well-being.

2. **Wellness Travel: A Journey of Gut Discovery**

Wellness travel becomes a celebration of both adventure and gut health. Embarking on wellness retreats worldwide immerse themselves in diverse practices—from Japanese onsen traditions to Nordic spa experiences. The journey becomes a tapestry of cultural wellness practices, each contributing to the holistic landscape of gut health.

Education and Awareness

1. **International Gut Health Symposium: Bridging Cultures and Sciences**

Organizing an international gut health symposium brings together experts from various cultures to share knowledge. Attending such symposiums gain insights into the latest research and traditional practices, fostering a global dialogue on gut health. The symposium becomes a nexus where science and cultural wisdom converge, celebrating the richness of knowledge available for cultivating gut well-being.

2. Online Gut Health Communities: Connecting Across Continents

Participating in online gut health communities offers a digital avenue to celebrate diversity. Engaging in discussions with individuals from around the globe, sharing experiences and insights, becomes a virtual celebration of the collective journey toward gut well-being. These online spaces become bridges that connect cultures and foster a sense of global community.

This global exploration invites us to embrace the wisdom of diverse traditions. From the probiotic allure of fermented foods to the nutrient-rich embrace of plant-based diets, each cultural thread contributes to the resilience and harmony of the gut. As men navigate the golden years, this chapter is a testament to the time-honored practices that transcend borders, uniting humanity in the collective pursuit of longevity and well-being.

Chapter Fifteen

Nature Therapy for Gut Wellness - Healing in the Outdoors

As men gracefully navigate the golden years, the quest for holistic well-being takes center stage. We will look into the transformative realm of nature therapy, illuminating the profound connection between time spent outdoors and gut wellness. In a world where technological advancements often dominate daily life, the healing power of nature emerges as a timeless ally, offering a sanctuary for both body and soul. While exploring the symbiotic relationship between nature therapy and gut health, unveiling the therapeutic benefits of immersing oneself in the great outdoors.

With its boundless beauty and intrinsic harmony, nature has long been recognized as a powerful source of healing for the human body, mind, and soul. Seeking holistic well-being can tap into nature's profound therapeutic effects. From reducing stress and enhancing cognitive function to fostering emotional resilience, the healing power of nature unfolds across various dimensions, creating a tapestry of wellness that transcends time and age.

The natural world, with its serene landscapes and rhythmic patterns, provides a respite from the demands of modern life. Iimmersing themselves in natural settings, witness a remarkable reduction in stress levels. The gentle rustle of leaves, the soothing sounds of flowing water, and the fresh scent of the outdoors orchestrate a symphony of serenity that allows the body to unwind. Nature acts as a natural stress buster, lowering cortisol levels and promoting a sense of calm that ripples the entire being.

Research has shown that exposure to nature, whether through walks in the park or simply spending time in green spaces, activates the parasympathetic nervous system. This activation counterbalances the fight-or-flight response, promoting relaxation and mitigating the negative impact of chronic stress. Navigating the complexities of life, nature becomes a gentle yet potent ally in the quest for emotional equilibrium.

Enhanced Cognitive Function: Nature's Impact on the Mind

The healing power of nature extends to cognitive well-being, offering a natural boost for mental clarity and focus. Time spent outdoors has been associated with improved concentration, heightened creativity, and enhanced problem-solving skills. Nature's ability to rejuvenate cognitive function is particularly relevant in an age where mental sharpness and acuity are valued components of a fulfilling life.

A phenomenon known as the "restorative effect of nature" suggests that exposure to natural environments allows the brain to recover from mental fatigue more effectively than in urban settings. Whether a stroll in a botanical garden, a hike through the woods, or simply gazing at a scenic landscape, nature provides a cognitive reset that invigorates the mind. Embracing nature's classrooms often have sharper mental faculties and a heightened sense of well-being.

Emotional Resilience: Nature's Elixir for the Soul

The healing power of nature extends beyond the physical and cognitive realms, reaching into the depths of emotional resilience. Natural settings have positively impacted mood and emotional well-being, offering solace in times of challenge or transition. Navigating the intricacies of life's journey find solace in the embrace of nature—a healing elixir for the soul.

Studies suggest that exposure to nature is linked to reductions in symptoms of depression and anxiety. Natural environments' visual and sensory richness prompts the brain to release mood-enhancing neurotransmitters, such as serotonin and dopamine. Additionally, the concept of "biophilia" asserts that humans have an innate connection to nature, underscoring the emotional sustenance derived from the natural world.

Physical Restoration: Nature's Gentle Remedy

Nature's healing touch extends to the physical body, offering a gentle remedy for various ailments and promoting overall well-being. Experiencing the outdoors becomes a form of physical restoration, contributing to improved immune function, better cardiovascular health, and enhanced recovery from illness or fatigue.

Natural settings provide opportunities for physical activity, whether through invigorating hikes, leisurely walks, or outdoor exercises. This movement contributes to improved circulation, muscle strength, and joint flexibility. Moreover, exposure to natural sunlight stimulates vitamin D production, which is crucial for bone health and overall immune function.

Connection and Social Well-Being: Nature as a Bonding Agent

Nature is a bonding agent, fostering individual connections and nurturing social well-being. Engaging in outdoor activities with friends, family, or community members becomes a pathway to meaningful connections. Whether it's a shared hike, a picnic in the park, or a leisurely day at the beach, nature provides the backdrop for shared experiences and strengthened social bonds.

Research indicates that spending time in nature can enhance prosocial behavior and improve interpersonal relationships. Nature's ability to facilitate positive social interactions contributes to emotional support and a sense of belonging, both crucial components of overall well-being, especially in the later stages of life.

Mindful Presence and Spiritual Connection: Communion with the Divine

Nature can facilitate mindful presence, inviting men you to immerse themselves in the present moment and connect with a deeper sense of purpose. Whether it's the awe-inspiring beauty of a mountain range, the rhythmic lull of ocean waves, or the quiet rustle of leaves in a forest, nature becomes a canvas for spiritual connection and contemplation.

For many individuals, nature serves as a conduit for a spiritual experience, offering a space for reflection, gratitude, and a sense of awe. Exploring the spiritual dimensions of nature often find themselves rejuvenated by a profound connection to something larger than themselves. This connection transcends the boundaries of time and space.

Biophilic Design: Bringing Nature Indoors

Recognizing the healing power of nature, architects and designers have embraced the concept of biophilic design—a practice that integrates natural elements into indoor spaces. Cultivating indoor environments infused with elements of nature experience the benefits of biophilia without necessarily venturing outdoors.

Biophilic design incorporates elements such as indoor plants, natural light, and materials inspired by nature to create spaces that mimic the outdoors. This approach has been associated with improved mood, reduced stress, and increased overall well-being. For those unable to spend extensive time in natural settings, biophilic design offers a harmonious alternative, bringing the healing essence of nature into daily living spaces.

The Therapeutic Landscape

1. **Forest Bathing: Immersion in the Healing Forest**

   Originating from Japan, Shinrin-yoku, or forest bathing, epitomizes the essence of nature therapy. Find solace in the embrace of lush greenery, immersing themselves in the healing compounds released by trees. The therapeutic landscape of the forest becomes a sanctuary for gut health, as phytoncides and other bioactive substances positively influence the gut microbiome.

2. **Coastal Retreats: The Salubrious Sea Connection**

   Coastal environments offer unique nature therapy with their rhythmic waves and refreshing sea air. Experiencing coastal retreats engage in a sensory journey beyond the shoreline. The sea, a source of diverse minerals and beneficial microbes, becomes a conduit for nurturing gut health. The chapter navigates the benefits of coastal environments, unveiling the holistic impact on both body and gut.

3. **Mountainous Heights: Elevating Gut Resilience**

Scaling mountainous heights through hiking or simply taking in panoramic views becomes a metaphorical ascent toward gut resilience. The crisp mountain air, enriched with phytochemicals and elevated oxygen levels, stimulates the gut's vitality. Conquering peaks to discover the interconnectedness between the elevated terrains and the thriving landscape of their gut microbiota.

Nature's Impact on Mental Well-Being

1. **Stress Reduction: Green Escapes for Relaxation**

   Chronic stress, a pervasive aspect of modern life, adversely affects gut health. Nature therapy emerges as a potent antidote, offering green escapes that induce a sense of tranquility. Engaging in stress-reducing activities amidst nature witness the transformative effects on their gut microbiota, fostering an environment conducive to balance and harmony.

2. **Mood Enhancement: Sunlight and Serotonin Connection**

Sunlight, a natural source of vitamin D, plays a crucial role in mood regulation. Basking in the sunlight during nature therapy sessions boost their vitamin D levels and enhance serotonin production. This mood-enhancing cascade reverberates through the gut-brain axis, contributing to emotional well-being and gut equilibrium.

3. **Mindful Presence: Meditation in Natural Settings**

Nature provides a canvas for mindfulness, inviting you to practice meditation amidst its serene landscapes. Mindful presence in natural settings becomes a conduit for gut health, as the meditative state positively influences the gut-brain axis.

Seasonal Dynamics and Gut Resilience

1. **Seasonal Eating: A Dance with Nature's Rhythms**

Aligning dietary choices with the seasons becomes an integral component of nature therapy. Adopting seasonal eating patterns synchronize their gut health with nature's rhythms. The chapter navigates the benefits of consuming seasonal, locally sourced foods, exploring the impact on gut diversity and overall resilience.

2. Biorhythms and Gut Microbiota: A Synchronized Symphony

Nature operates on rhythmic cycles, and so does the human body. Embracing nature therapy delve into the interconnected biorhythms that govern gut microbiota. From circadian rhythms to seasonal fluctuations, the chapter unravels the synchronized symphony between the body's internal clock and the dynamic landscape of the gut microbiome.

Immersion in Natural Elements

1. Earthing: Grounding for Gut Vitality

The practice of earthing, connecting with the Earth's surface, takes center stage in nature therapy. You should try to ground yourself on natural terrains and experience a surge in antioxidants and the transfer of electrons from the Earth to the body.

2. Hydrotherapy: Aquatic Healing for Gut Harmony

Hydrotherapy, incorporating water as a healing element, becomes a cornerstone of nature therapy. Immersing yourself in natural bodies of water, embrace the therapeutic benefits that extend to the gut.

Biophilia and Gut Harmony

1. **Biophilic Design: Nature's Influence in Living Spaces**

The concept of biophilic design, integrating natural elements into living spaces, extends the benefits of nature therapy beyond outdoor excursions. When you are urrounded by biophilic environments at home or in communal spaces, experience the ongoing influence of nature on gut health.

2. **Companion Animals: Furry Friends and Gut Microbiota**

The presence of companion animals, often considered family members, contributes to the biophilic connection. If you are sharing your lives with furry friends you tend to witness the impact of pet companionship on gut microbiota diversity.

Cultural Perspectives on Nature Therapy

1. **Shinrin-yoku in Japan: Cultivating Forest Wisdom**

   Nature therapy finds its roots in Shinrin-yoku, the Japanese practice of forest bathing. Exploring Shinrin-yoku draw inspiration from Japan's cultural reverence for nature.

2. **Nordic Friluftsliv: Embracing Outdoor Living**

   The Nordic concept of Friluftsliv, embracing outdoor living, becomes a beacon for seeking a harmonious connection with nature. The chapter navigates the Nordic philosophy and its implications for gut wellness. From outdoor activities to embracing natural elements, Friluftsliv becomes a cultural guide to fostering gut resilience in the Nordic tradition.

3. **Ancient Practices in Indigenous Cultures: Nature's Wisdom**

   Indigenous cultures worldwide have preserved ancient practices that honor the wisdom of nature. Immersing yourslef in the teachings of indigenous communities, discover profound insights into the relationship between cultural heritage, nature therapy, and gut well-being. The chapter explores the diverse practices of indigenous cultures and their relevance to the pursuit of gut resilience.

Nature Therapy as a Lifestyle

1. **Gardening for Gut Health: Tending to Nature's Bounty**

   Cultivating a garden becomes a therapeutic endeavor, intertwining nature therapy with nurturing life. Men planting, tending, and harvesting from their gardens witness the transformative impact on gut health.

2. **Outdoor Fitness: Exercise in Nature's Gym**

   Traditional fitness regimens meet the rejuvenating embrace of nature in outdoor exercise. Engaging in activities like hiking, cycling, or outdoor yoga infuse your workouts with the therapeutic elements of nature.

Conclusion

In the expansive journey of life, the quest for well-being is a constant companion, guiding individuals through diverse landscapes of experiences, challenges, and triumphs. As men gracefully navigate the intricacies of aging, the significance of gut health emerges as a linchpin to overall vitality. This comprehensive guide, meticulously tailored for men over 60, has traversed the vast terrain of gut well-being—a journey that spans scientific insights, cultural wisdom, and practical lifestyle applications. As we conclude this exploration, it is not merely an endpoint but an invitation to embrace a lifetime of gut well-being. This holistic approach transcends individual chapters to weave a tapestry of enduring health and vibrancy.

The journey embarked upon in this guide is one of reflection and understanding—a voyage into the intricate ecosystems that reside within the human gut. From the evolution of gut health practices to the delicate balance of the microbiota, each chapter has unveiled layers of knowledge to empower you with insights to nurture their gut well-being. Reflecting on this journey involves acknowledging the dynamic interplay between lifestyle, diet, and the countless microorganisms that call the gut home.

Gut health is not a fleeting concern but a lifelong companion intricately woven into the fabric of daily life. As men traverse the various seasons of existence, from the exuberance of youth to the wisdom of older age, the gut remains a constant steward of well-being. Embracing this perspective invites a shift from viewing gut health as a reactive endeavor to understanding it as a proactive, lifelong commitment that evolves and adapts to the changing landscapes of life.

The guide has seamlessly integrated scientific knowledge with cultural wisdom, recognizing that the pursuit of gut well-being encompasses empirical evidence and timeless insights from diverse cultures. Scientific revelations about the gut-brain axis, the role of microbiota in immune function, and the impact of lifestyle on gut diversity converge with cultural practices, such as fermented foods, herbal remedies, and mindful nutrition. The culmination of these insights forms a harmonious symphony—a blend of evidence-based understanding, and the nuanced wisdom passed down through generations.

In recognizing the unique needs and considerations, the guide has unfolded a series of tailored approaches. From understanding the impact of aging on gut health to navigating common digestive challenges, each chapter has served as a compass, guiding men through the intricacies of maintaining gut well-being in this specific stage of life. The tailored nature of the guide ensures that the insights provided resonate with the experiences and aspirations of men navigating the golden years.

At the heart of gut well-being lies the integration of lifestyle practices—a harmonious interplay of nutrition, physical activity, mental well-being, and the profound connection with nature. The chapters dedicated to mindful nutrition, physical exercise, and the healing power of nature have provided a roadmap for you to cultivate a holistic lifestyle that nurtures the gut from multiple angles. Recognizing that gut health is not isolated but intertwined with the choices made in daily life forms the cornerstone of a resilient approach.

As you embrace a lifetime of gut well-being, they acknowledge the gut as the epicenter of health. This dynamic nexus influences digestion, immune function, mental clarity, and overall vitality. The gut-brain axis, the microbial tapestry within, and the intricate dance between the gut and other bodily systems underscore the central role of gut health in sustaining a high quality of life.

Beyond immediate well-being, the guide has explored strategies for longevity and vitality—unraveling the profound interconnection between gut health and the prevention of age-related illnesses. The chapters on chronic diseases, the gut-brain connection, and physical performance illuminate the potential of maintaining a healthy gut microbiota as a potent preventive measure against the challenges that can accompany aging.

The guide has celebrated the diversity of gut health practices worldwide, recognizing that cultural wisdom contributes to a global tapestry of well-being. From Shinrin-yoku in Japan to Nordic friluftsliv, from Ayurvedic cooking to African superfoods, each cultural insight becomes a brushstroke in painting a rich, global approach to gut health. By embracing this diversity, you become guardians of your health and stewards of a collective, global legacy.

As the guide concludes, the spotlight turns to nature therapy—a lifelong ally in pursuing gut well-being. The healing power of nature, explored in depth in the final chapters, invites you to immerse yourself in the refreshing embrace of the outdoors. Nature becomes a sanctuary where stress dissipates, cognitive function flourishes, and emotional resilience thrives. The call to nature is not a transient chapter but a perpetual invitation—a timeless journey where the healing elements of the natural world continue to unfold.

The conclusion of this guide is not an endpoint but a recognition that gut well-being is woven into the very fabric of daily choices. Every meal, every step taken, every moment of mindfulness, and every connection with nature becomes a thread in the tapestry of health. As you move forward, you do so with an enriched understanding, a profound connection to the cultural legacy of well-being, and an enduring commitment to the lifelong journey of gut health.

In the grand mosaic of life, gut well-being is not a destination—it is a journey, a continuous dance with the rhythms of existence. Armed with the insights garnered from this guide, embark on this journey with a sense of empowerment, a spirit of curiosity, and a commitment to nurturing your gut health across the vast expanse of a lifetime. May this journey be one of enduring vitality, holistic well-being, and the joyous celebration of a lifetime of gut health.

References

García-Montero, C., Fraile-Martínez, O., Gómez-Lahoz, A. M., Pekarek, L., Castellanos, A. J., Noguerales-Fraguas, F., Coca, S., Guijarro, L. G., García-Honduvilla, N., Asúnsolo, A., Sanchez-Trujillo, L., Lahera, G., Bujan, J., Monserrat, J., Álvarez-Mon, M., Álvarez-Mon, M. A., & Ortega, M. A. (2021). Nutritional components in western diet versus Mediterranean diet at the gut microbiota–immune system interplay. implications for health and disease. Nutrients, 13(2), 699. https://doi.org/10.3390/nu13020699

Anderson, S. (2022, July 8). Over 40% of Americans are deficient in this vitamin: Here are the symptoms to look out for. Rupa Health. https://www.rupahealth.com/post/what-causes-vitamin-d-deficiency

Christie, J. (2022, December 13). 95% of American's aren't getting enough fiber: How many grams should we be consuming per day? Rupa Health. https://www.rupahealth.com/post/95-of-americans-arentgetting-enough-fiber-how-many-grams-of-fiber-should-we-be-consuming-per-day

Ghosh, S., & Iacucci, M. (2021). Diverse immune effects of bovine colostrum and benefits in human health and disease. Nutrients, 13(11), 3798. https://doi.org/10.3390/nu13113798

Wiertsema, S. P., van Bergenhenegouwen, J., Garssen, J., & Knippels, L. M. (2021). The interplay between the gut microbiome and the immune system in the context of infectious diseases throughout life and the role of Nutrition in Optimizing Treatment Strategies. Nutrients, 13(3), 886. https://doi.org/10.3390/nu13030886

Wastyk, H. C., Fragiadakis, G. K., Perelman, D., Dahan, D., Merrill, B. D., Yu, F. B., Topf, M., Gonzalez, C. G., Van Treuren, W., Han, S., Robinson, J. L., Elias, J. E., Sonnenburg, E. D., Gardner, C. D., & Sonnenburg, J. L. (2021). Gut-microbiota-targeted diets modulate human immune status. Cell, 184(16). https://doi.org/10.1016/j.cell.2021.06.019

Printed in Dunstable, United Kingdom